Plankton Culture Manual

Sixth Edition
By

Frank H. Hoff and Terry W. Snell
© 1987

Published by

Florida Aqua Farms, Inc.
33418 Old Saint Joe Road
Dade City, Florida 33525

10 Digit ISBN# 0-9662960-4-4
13 Digit ISBN 978-0-9662960-4-4

First Edition, February 1987
Second Edition, June 1989
Third Edition, May 1993
Fourth Edition, December 1997
Fifth Edition, July 1999 & May 2001
Sixth Edition, 1st Printing, March 2004
Sixth Edition, 2nd Printing, February 2007
Sixth Edition, 3rd Printing, November 2008
Sixth Edition, 4th Printing, June 2014

D1430174

This manual is dedicated to our wives and
families
Nancy Jo, Autumn, Dustin, and Tony Hoff
Sandra and Sara Snell

We want to recognize Barry Rosen Ph.D and Roger W. Rottmann
for their contribution to portions of this manual.
A very special thank you is extended to our editor Jeff Neslen for
his excellent help.

Preface

Over the twenty years of publication the purpose of our manual has always been to provide aquarists, aquaculturists and students with techniques and confidence to culture plankton for use live foods for aquatic animals. The approach we have taken is to provide methods on a variety of scales ranging from home hobbyists to large commercial operations. The Plankton Culture Manual serves as a companion to another manual by Frank Hoff, Conditioning, Spawning and Rearing of Fish With Emphasis on Marine Clownfish. The Plankton Culture Manual provides the basis and principles of live food culture while the companion book provides theory and practice of conditioning, spawning and utilization of live feeds in larval rearing, juvenile gro-wout, and adult broodstock fish. Methods of plankton culture described here are the result of cumulative experiences during our many years in aquaculture mixed with numerous published works and direct communication from aquaculture scientists and hobbyists throughout the world. Using advanced computer data bases and internet of scientific and hobbyist literature, we have culled out significant papers that have made substantial contributions in live food culture. Literature citations in this manual provide readers access to a vast selection of scientific and hobby publications that demonstrate the dynamic changes occurring in live food culture.

Through the sixth editions of this manual, we have continued to update, improved, clarified, and provided more detail about the methods used or in use. In addition, we have provided more rationale of why certain methods are preferred over others. Ultimately we hope this book provides readers with the guidance they need to provide live food in the culture of aquatic animals for fun and profit. Our hope is that this useful manual will continue to be improved and grow in the years to follow. Happy culturing!

Frank Hoff and Terry Snell,
June 2008

About the Authors

Frank H. Hoff - Frank Hoff has over 42 years experience in aquaculture. He served as head biologist of the Aquaculture Division for the State of Florida, Marine Research Laboratory then moved into the private sector as founder of Instant Ocean Hatcheries, pioneering the use of closed, recirculating, artificial seawater systems to raise clownfish. Earned a Bachelor degree in Zoology and a Masters degree in Marine Science at the University of South Florida. Founded and currently an owner of Florida Aqua Farms, Inc. started in 1984 and a past owner of Aquaculture Supply started in 1987. Consultant and board member for Ocean Reefs and Aquariums (ORA), the largest marine tropical fish farm in the world, Serves as a board member of the Florida Aquaculture Association.

Terry W. Snell - Dr. Snell received his Ph.D. in biology from the University of South Florida and has been a college professor for the past 32 years. He currently teaches ecology, genetics, population biology, and aquatic toxicology at Georgia Tech in Atlanta. His research focuses on the evolution of reproductive isolation in rotifers and copepods, ecotoxicology, and plankton culture. He has been a consultant to the marine and freshwater aquaculture industry for many years.

TABLE OF CONTENTS

Chapter I - AQUATIC FOOD WEBS

Plankton is an encompassing word for passively drifting or weakly swimming organisms found in marine and freshwater environments. Members of this group range in size from microscopic, single-celled plants to jellyfish measuring up to six feet across, including eggs and larval stages of more complex animals such as fish and corals. The primary focus of this manual is the culture of microscopic forms of phytoplankton (plant plankton) and specific zooplankton (animal plankton) commonly used in aquaculture as food for larval fish and invertebrates.

Our goal is to clearly describe the technical aspects of culturing phytoplankton and zooplankton so that anyone who carefully follows these instructions can grow all the microalgae, rotifers, ciliates, *Artemia,* daphnids and clam or oyster veligers they desire. Information is presented at a level that should be comprehensible to those with minimal experience in aquaculture. The descriptions are simple, but with sufficient detail so that most people can succeed on their first try. We have provided technical data and illustrations to support the rationale behind our instructions. Our overall intention is not only to provide technology for growing live food, but to promote a more thorough understanding of aquatic environments and the ecological relationships of food webs.

The methods described have been employed successfully in commercial aquaculture for many years. We have simplified standard methods for small scale cultures that can be maintained at home by hobbyists and provided new protocols for large scale commercial cultures. We have selected methods that do not require stringently controlled temperatures, salinities, or light levels and can be accomplished in non-sterile environments.

The objective in producing live food is to culture billions of microscopic algae cells to feed millions of zooplankton which are then fed to thousands of larval fish and invertebrates. This principle is known in ecology as a food web (Figure 1.1). These interrelationships usually keep populations in balance in natural environments through positive and negative feedbacks. However, the food web in an aquarium or culture tank is incomplete. Aquarists try to overcome this deficiency by substituting flake foods, gelatin based mixtures, and frozen or freeze-dried foods. Regardless of how careful one is in feeding non-living or processed foods, problems usually arise due to nutritional deficiencies or water quality degradation. Experience in the culture of fish and aquatic invertebrates has shown that better growth and survival are obtained when utilizing live foods exclusively or in combination with inert foods.

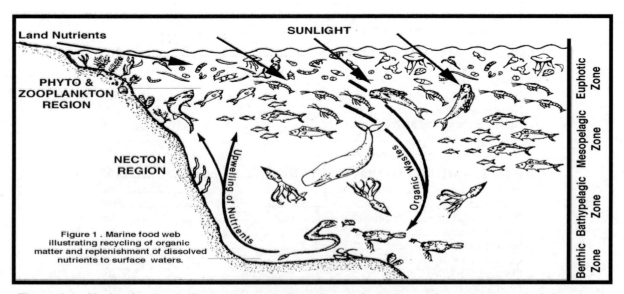

Figure 1 . Marine food web illustrating recycling of organic matter and replenishment of dissolved nutrients to surface waters.

Figure 1.1 - Marine food web illustrating recycling of organic matter and replenishment of dissolved nutrients of surface waters.

The Role of Phytoplankton

Biomass Production

Phytoplankton are graded according to size. Macroplankton are larger than 1 mm, microplankton are less then 1 mm, nanoplankton are 5-60 micrometers (μm) and ultraplankton which are less than 5 μm. Our concerns are phytoplankton within the nanoplankton and ultraplankton range of species. Phytoplankton or microalgae, are the synthesizers of organic matter in aquatic environments. Microalgae utilize photosynthesis to construct complex carbon molecules like the more familiar land plants. Dissolved nutrients from various sources are absorbed by microalgae and, in the presence of light, combined into more complex molecules necessary for survival. Many microalgae are heterotrophic and reproduce in the dark providing a carbon source is available while others are mixotrophic that can reproduce in light or dark conditions. Microalgae are major contributors to the production of biomass in oceans, estuaries, lakes, and reservoirs. Although small, the combined effects of billions of these cells provide most of the plant material consumed by animals higher in the food web. The entire food web receives its energy from the biomolecules synthesized by these microscopic plants. For example, production on the surface of the open ocean is about 50 grams or approximately one cup of carbon per square meter per year, almost all of which is due to phytoplankton photosynthesis (carbon atoms are the building blocks of all complex organic molecules). Annual production of all plant life is estimated at 100 x 10$_9$ metric tons of fixed carbon of which phytoplankton is a major contributor (Boney 1983). With this kind of productivity, it is no wonder the open ocean and phytoplankton within are referred to as the pastures of the sea.

Nutrient recycling

In addition to supplying food for animals, phytoplankton play a central role in nutrient cycling in aquatic habitats (Figure 1.2). Microalgae absorb primary nutrients such as ammonia, urea, nitrates, phosphates, and potassium, as well as metals such as iron, copper, manganese, zinc, molybdenum, and vanadium. Certain vitamins such as B_{12}, thiamin, and biotin also have been found to be essential or beneficial to many algae species and are selectively absorbed or produced by some species. Removal and alteration of nutrients by phytoplankton in the ocean is not permanent, since nutrients are slowly being regenerated as microalgae cells die and decompose. Other sources of dissolved nutrients include terrestrial runoff, rain, and coastal upwelling. In marine and freshwater aquaria or culture tanks, phytoplankton help maintain water quality through their removal of excess nutrients and the regulation of pH. In effect, each algae cell is a living biofilter. For more details about the role of algae in aquatic ecosystems see Round (1981) or Bold & Wynne (1985).

The Role of Zooplankton

Primary grazers

Zooplankton are microscopic or somewhat larger grazing animals in aquatic environments. Many zooplankton feed on phytoplankton and form the second major link of the food web. Even though "plankton" means passively floating or drifting, some zooplankton are actually fairly strong swimmers. Zooplankton swim or drift through clouds of phytoplankton, grazing on the cells that contact their feeding appendages. Just any microalgae cell won't do, however. Many zooplankton are very choosy about the microalgae they eat. Even though small, the collective effect of zooplankton feeding greatly reduces phytoplankton abundance at certain times of the year. The yearly plankton cycle consists of various phytoplankton species blooming in response to a particular sequence of environmental triggers including, changes in temperature, salinity, light duration, light intensity, and nutrient availability. After a short time lag of several days or weeks, zooplankton populations respond to specific environmental triggers including the increase in phytoplankton abundance. Phytoplankton and zooplankton blooms are usually sequential, with one set of species blooming during a certain time followed by another set of species in a seasonal cycle. Phytoplankton and zooplankton populations are therefore intimately linked in a continuous environmentally triggered cycle of bloom and decline that has evolved and persisted throughout millions of years of evolution.

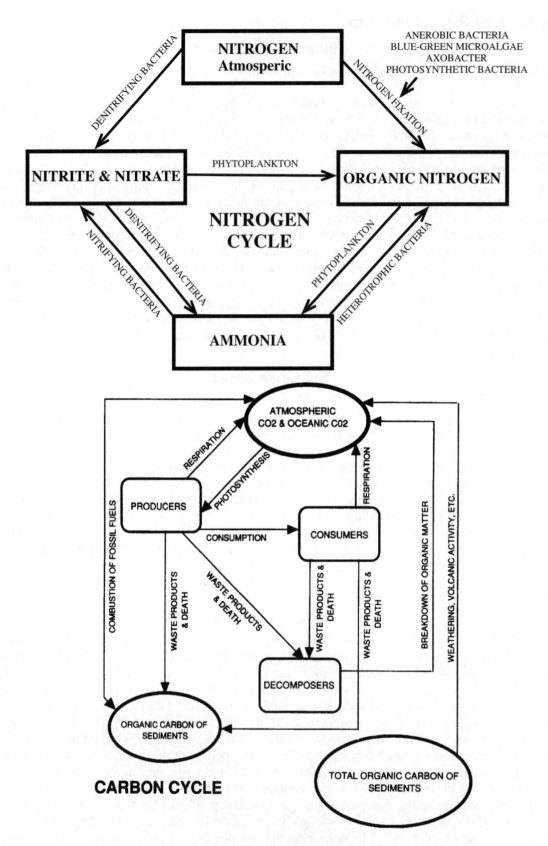

Figure 1.2 - Nutrient recycling pathways, nitrogen cycle (top) and carbon cycle (bottom).

Chapter 2 - UTILIZATION OF PLANKTON

Rules When Using Live Foods

Spotte (1971) provided a very good overview and guidelines for utilizing live foods which still applies to current thoughts. The following is a synopsis of this article. Over the years there is much controversy about whether or not it is advantageous to feed live food organisms to captive aquatic animals. On one side there are those who believe live food to be substantially more nutritious, yielding faster and more even growth. Then there are believers that scientifically-blended, non-living mixture, of prepared foods can give equally good growth and development with fewer risks of parasitic invasion of culture systems. As is the case, the truth lies somewhere in between.

Specific foods may be necessary to pull a cultured animal through one or more delicate developmental stages where its growth is critical within a short period of time. Many larval fishes and invertebrates will not take anything but live foods during such periods. Newly-hatched pelagic larvae often require a live food smaller than new-hatch brine shrimp. Plaice have been successfully reared through the first four days on just nauplii of barnacles followed by a mixture with brine shrimp and barnacle larvae. In recent years live rotifers have become the most acceptable small initial food.

In most cases live foods are indeed superior from a nutritional standpoint. On the other hand they may also be dangerous, particularly when taken directly from the wild. The danger is magnified when the collected food organisms form a part of the natural diet of the animals under culture. If the evolutionary relationship between the two has been lengthy, the chances are that at some point a parasite has made it a threesome. Often the parasite is highly species-specific, afflicting only a single host, or a series of hosts, in a food chain. If more that one host is required for the parasite to complete its life-cycle, the last one is called the definitive host. All those before are intermediate hosts.

There is immediate danger of parasite transfer when the live food organism is a step lower in the food chain than the animals being cultured. Transfer may take place internally when the definitive host eats the intermediate host (along with the live food) and ingests the parasite. Or transfer may be external when the parasite leaves the intermediate host.

In the wild parasites take a toll of their hosts, but never to a point of eliminating them completely since this would also destroy the parasite specie. In captivity the situation can be different. There is no such thing as a completely captive ecosystem. Checks and balances that normally serve as tempering mechanism in the wild state are either distorted or do not exist at all under captive conditions. Partial isolation from the normal ecosystem can suddenly upset the equilibrium between host and parasite and tip the scales heavily in favor of the parasite. Normal immunities acquired by a host species do not always hold up during the stress of confinement. To reduce or eliminate the possibility of a problem there are specific general rules to consider.

1) The thread from species-specific parasites can be reduced by not feeding cultured animals live food organisms from the same food chain.

Most ectoparasitic protozoans and helminths are adversely affected by changes in the salinity of the external medium. True marine forms cannot osmoregulate and are unable to survive for extended periods in freshwater. The reverse is also true and freshwater species that soon become desiccated and die when placed in seawater. This is only a generality and should be accepted as such. It becomes even less valid where pathogenic bacteria and viruses are concerned.

2) Live freshwater food organisms can be fed to marine animals with reduced risk of parasitic infection or infestation. The reverse is also true seawater species.

Herald and Dempster (1955) showed that brine shrimp maintained in copper-treated seawater could concentrate up to 2.0 ppm copper in their tissues. Even levels far less were enough to kill small coral reef fishes when they were fed on the contaminated shrimp. Daphnia and brine shrimp may also pick up zinc ions from galvanized holding containers. Brass and copper fittings on the containers can be another source of potential contamination.

3) Holding and culture facilities for live food organisms should be chemically inert, preferably polypropylene, cured polyester or epoxy or glass.

Hiyama and Singh (1966) used a radioisotope (S35) as a tracer to measure the absorption rates of different types of live prey by carnivorous fishes. Live guppies, brine shrimp, and daphnia were all found to be excellent foods. So were fresh samples of muscle tissue from bivalve mollusks, gobies, and goldfish. Midge larvae, however, proved to be a poor food for marine fishes, although they were more than adequate as food for freshwater species.

4) Absorption rates of a given species of live prey vary with the species of the predator. Basically, there is no universally "ideal" live food organism.

Phytoplankton

Primary food source

Nutritionally, microalgae (phytoplankton) are a source of micronutrients, vitamins, oils, and trace elements for aquatic communities. They are rich sources of macronutrients, protein, carbohydrates, and especially specific essential fatty acids (See Tables 4.1 & 4.2). Microalgae provide essential pigments such as astaxanthin, zeaxanthin, chlorophyll, and phycocyanin which enhance coloration and health in fish and invertebrates. Iodine is an example of a primary trace element supplied by algae and is essential to virtually all living organisms. As a food source microalgae exceeds all others.

Spirulina is an example of a commercially important phytoplankton that is native to alkaline lakes throughout the world. Direct utilization of this microalgae as a human food source is recorded by Maya Indians in Central America has been practiced for over 600 years. It was used in cakes or soup. Current natives of Chad in Africa use *Spirulina* in cakes and soups daily. Recently *Spirulina* has become a popular human health food and is available in powder, pellet, or utilized as a additive in animal foods and aquarium flake foods. In Africa, *Spirulina* supports huge flocks of flamingos and soda Tilapia (cichlid fish). Dried *Spirulina* is 65-68% protein and is often used as a supplemental diet for culturing fish, adult *Artemia* and other invertebrates. An acre of *Spirulina* can produce approximately 10 long tons of proteins per year compared to wheat at only 0.16 long tons and beef 0.016 long tons. Annual world production of *Spirulina* is greater than 900 tons dry weight, with over 100 tons farmed in the USA. At least 40 food manufacturers in the USA are currently utilizing *Spirulina* as food supplements or ingredients. Products include high-protein drinks, pastas, and snacks (Duxbury 1989). For more information regarding the utilization of microalgae refer to Kinne (1977) and Goldman (1979).

Chlorella and *Scenedesmus* are other unicellular microalgae that have become popular as human and animal feed sources. Under semi-controlled growing conditions the yield of dried *Chlorella* can range from 20-60 tons/acre/year with an average of 40 tons. Compared to other conventional terrestrial crops is extraordinary. Averages in tons/year/acre of soybean, 0.5 tons, rice 1.8 tons and alfalfa grass 5 tons. The nutritional value of one pound of *Chlorella* paste corresponds to 2 pounds of soybean flour. Wet *Chlorella* is 80% digestible and dried powder about 65%. A one acre, 3 foot deep pond under continuous culture could conceivably provide enough nutrition to feed 1000 cows per year (Hills and Nakamura 1978).

Various live species of microalgae are used as food in the culture of filter-feeding mollusks such as clams, mussels, oysters, and scallops. Combinations of algae species are sequentially utilized in the larval culture of marine shrimp. More recently, marine aquarist have found microalgae to be beneficial in maintaining corals, sponges, barnacles, tube-worms, sea squirts, and other filter-feeding invertebrates. Utilization of various species of microalgae are listed in Table 2.1 (modified from Okauchi, 1991, and Rosen, 1990).

Source of various products

Microalgae are capable of rapid population growth and play a major role in nutrient recycling. In addition, they help balance the pH of aquatic systems by removing excess carbon dioxide and adding oxygen. They have been cultured to produce oils, chemicals, pharmaceuticals, and polysaccharides, as well as for eutrophication control. Sewage treatment and water recovery facilities have utilized microalgae for many years to remove primary nutrients. Instant Ocean Hatcheries, a closed-system marine tropical fish hatchery, utilized an outside recovery system consisting of six, ten thousand gallon (38 cubic meters) cement ponds. These were provided with heavy aeration which supported high densities of micro- and macroalgae. Nitrates as high as 30-40 ppm (parts per million) were reduced to 0-5 ppm in three to four months. This inland marine hatchery was able to consistently recover and reuse 70% of the total water volume of 110,000 gallons.

Algal Species	Aquatic Animals Cultured
Isochysis galbana - golden brown motile, size 4-8 μ	rotifers, clams, oysters, conch, sea cucumbers, sea hares (saltwater use)
Nannochloropsis oculata - golden brown, non-motile, size 4-6 μ	rotifers, brine shrimp, daphnia. monia, marine shrimp (fresh & saltwater use)
Tetraselmus sp. - green, motile, size 9-10 x 12-14 μ	rotifers, marine shrimp (saltwater use)
Chlorella vulgaris - green, non-motile, size 2-10 μ	rotifers, protozoans (fresh & saltwater use)
Nitzchia sp. - diatom, non-motile	abalones, turbans (saltwater use)
Navicula sp. - diatom, non-motile	abalones, turbans (saltwater use)
Phaeodactylum tricornutum -diatom, motile, size 3-5 x 12-35 μ	spiny lobsters, clams, oysters (saltwater use)
Thallasiosira sp. - diatom, non-motile, size 11-14 x 14-17 μ	clams, oysters, scallops, larval shrimp (saltwater use)
Chaetoceros gracillis - diatom, non-motile, size 14 x 17 μ	clams, oysters, scallops, shrimp, sea urchins, conchs, sea cucumbers (saltwater)

Table 2.1 - Algae species used to raise various invertebrates.

Therapeutic effect

In addition to these well documented contributions of microalgae to aquaculture, there are some other less understood benefits. As water conditioners, microalgae have been shown to have a therapeutic, antibiotic-like effect on fish and other animals. These effects may be the result of compounds released by algae rather than what the algae cells remove from the water. *Spirulina* has been shown to enhance the immune system in fish, invertebrates, and chickens. The National Cancer

Institute announced the discovery of glycolipids in blue-green algae which are active against the AIDS virus. Scientists are now studying how to enhance glycolipid content during culture (Duxbury, 1989). Several large drug companies are currently culturing many species of microalgae in search of pharmaceutical ingredients.

Zooplankton

Utilization as primary live food

For years marine rotifers and *Artemia* nauplii have been utilized as the initial food for larval marine fish. More recently, they have been utilized commercially to rear larval marine shrimp and freshwater fish. Other zooplankton such as copepods, protozoans, and larval stages of urchins, oysters, and clams are also utilized, but thus far rotifers, *Daphnia,* and *Artemia* have proven the most effective.

Rotifers have a number of features that make them useful for aquaculture, including rapid reproduction, small size, slow swimming speed, high nutritional quality, and ease of culture. A thousand rotifers, fed properly, can multiply to over a million animals in five to seven days at 25° C (77° F). Rotifers are about the one-third the size of newly hatched *Artemia* nauplii. Body size of brachionid rotifers differs according to species and strain, but generally ranges from about 120-500 μm (micrometers) in length (.0047-.0196 inch) and about 100-200 μm in width (.0039-.0078 inch). The smallest newly hatched *Artemia* nauplii are about 400 μm long and 250 μm wide.

Rotifers currently are used worldwide to culture over 60 species of marine finfish, several species of freshwater fish, and 18 crustaceans (Nagata 1989, Hirayama and Hagiwara 1995). Production on just one farm in Japan can be as high as 2.5 tons of rotifers per year. These are utilized to raise 6.3 million larval red and black sea bream (12-16 mm long) and 4 million crabs, *Portunus trituberculatus* (Fukusho 1989). It has been calculated that in culturing red sea bream larvae over their 25 day larval rearing period, one larvae will consume 12,000 to 15,000 rotifers per day, and reach a length of 10 mm (0.39 of an inch).

Controlled live food source

The biochemical composition of rotifers and *Artemia* closely reflects what they eat. Rotifers and *Artemia* can be fed specific fatty acids, amino acids, vitamins, and antibiotics that in turn can be transferred to larval fish or invertebrates. For example, when rearing clownfish at Instant Ocean Hatcheries, there was an incident where vitamin B_{12} (cyanocobalamin) was omitted from the microalgae fertilizer. Since this particular algae species *(Pyramimonas)* apparently grew well without this trace nutrient, it was assumed to be unnecessary. Within several weeks, heavy larval clownfish mortalities were noted about three days after they were fed the B_{12} deficient rotifers. Apparently, the algae were B_{12} deficient and this deficiency was directly passed to the rotifers. The problem was almost immediately corrected when B_{12} was restored to the algae growth medium. Others have found that a B_{12} deficiency in rotifer cultures reduces rotifer growth and reproduction (see Rotifer Culture Section). How or in what form nutrients are passed from the algae to rotifers and fish larvae is not clearly understood. Nevertheless, the B_{12} example, clearly demonstrates the impact a nutritionally deficient alga can have on larval survival.

To insure high quality rotifers and *Artemia*, it is important to "pack" them with essential fatty acids (HUFA), vitamins, and other nutrients just before feeding them to larval fish and invertebrates. This can be accomplished by using high quality algae containing essential fatty acids or commercial enrichment products. Just because a certain microalgae grows well under your conditions, does not mean that it is sufficiently nutritious for culturing larvae. All microalgae are not alike. Even subspecies or strains of one algae species differ markedly in their HUFA composition (see Microalgae Chapter).

Chapter 3 - MICROALGAE IDENTIFICATION

By Barry Rosen, Ph.D.

Introduction

Proper taxonomic identification is important for producing algal feeds of consistent nutritional quality. In addition, cultures of microalgae may become contaminated with nuisance organisms of a less desirable strain. Correct taxonomic classification is required to recognize the contaminant and take appropriate action. Microalgae change in form, cell size, and motility during different parts of their life cycle or because of culture conditions. These changes can lead to improper species identification. Some microalgae species produce toxins that are harmful to other algae and to the animals that feed on them. Therefore, proper identification of desirable and unwanted microalgae species is essential for aquaculturists. Because of the great number of microalgae species and their difficulty of identification, we have limited our discussion only to species currently utilized in aquaculture. To identify other species of microalgae, consult Smith (1950), Prescott (1962), or Whitford and Schumacher (1969).

It is common for microalgae to become progressively smaller or distorted in shape with prolonged culturing. One species may resemble another due to salinity, light, or nutrient conditions. The morphology of microalgae is flexible and reflects their ability to adapt to a variety of environmental conditions. Algal morphology, therefore, is not the most reliable characteristic for classifying species. Unequivocal algal classification is based on pigment biochemistry and genetic similarities.

Microalgae are classified as plants by most biologists because they contain chlorophyll and have a cell wall like higher plants (Figure 3.1). Some biologists classify the motile forms with the Protozoans. Most recent attempts at classification unify all single-cell eukaryotic organisms and multicellular algae (including macroalgae) into the Kingdom Protista, a separate kingdom from both plants and animals. Under any classification scheme, species of microalgae are defined by their unique morphology and biochemistry, although morphological variations from a generalized description are common. Some genera have several species or well characterized strains that are used in aquaculture. Some strains are not visibly distinguishable with a common light microscope and biochemical techniques are required to separate them.

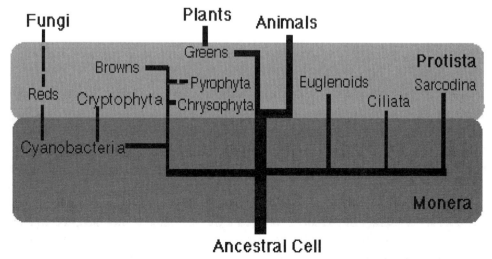

Figure 3.1. The phylogenetic relationships of algae.

The topics in this chapter are organized as follows: the proper use of the light microscope, a brief overview of each major taxonomic algal group, a detailed description and photomicrographs of several important algae, and nuisance organisms.

Observing Microalgae

Microscopes

A compound light microscope is the essential tool for proper algal identification. Each species has a specific size and shape that is used for identification and requires detailed observations of cell structure. In order to determine these characteristics, many of which are subtle, a compound microscope must be used to its fullest potential. In addition, the specimen should always be examined at the highest possible magnification to see as many cellular details as possible.

Microscope Illumination

Perhaps the most important aspect of the microscope is achieving proper illumination of the specimen. Many organisms are translucent and may be obscured if too much light is used. On the other hand, small organisms must be observed at high magnification (at least 100X), which requires more light. Two light sources are common with compound scopes: those that have a mirror which reflects light toward the specimen and those with built-in illuminators (usually a 6-50 watt incandescent bulb). If the microscope has a reflecting mirror, the first task is to position the mirror to reflect as much light as possible into the body of the microscope. If the eyepiece can be removed, you can look down the barrel of the microscope and center the incoming light.

If the microscope has a built-in illuminator, the contrast of the specimens viewed through the microscope can be enhanced by using a blue or green filter above the light

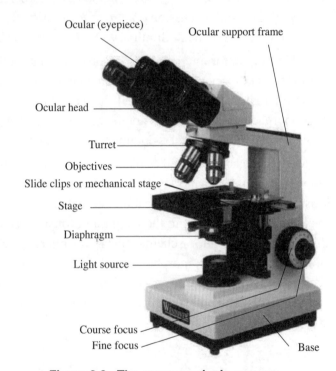

Figure 3.2. The compound microscope.

source. These filters increase the resolution by eliminating longer wavelengths of light that are emitted from incandescent bulbs.

Some kinds of microscopes have a condenser that focuses the light from the source before it passes through the specimen. The position of the condenser relative to the specimen is important and may be adjustable. The knob below the stage is used to make adjustments in order to focus the maximum amount of light on the specimen.

In addition to the condenser, microscopes have an iris diaphragm or a series of holes on a metal dish above the condenser and below the stage. This device is the key to improving the microscope image. By closing down the iris or using a smaller hole, especially on low power, stray light is removed which enhances contrast and visibility of the specimen. After the specimen is mounted, the condenser and iris positions should be sought which give the best image before attempting to identify algae. Also keep in mind that the iris may need to be adjusted with each change in magnification.

Lenses

The part of the microscope you look through is the ocular lens. Some microscopes are binocular, having two ocular lenses. Binocular microscopes are more comfortable, especially for extended observation, but are more expensive. If the microscope has two ocular lenses, adjust the distance between them until it is comfortable to examine the specimen and one image is present. Some ocular lenses can also be individually focused. If there are two ocular lenses and one lens can be focused, first focus the fixed lens on the specimen with the microscope focus knob, then focus the ocular lens to the same plane of focus.

The magnification of the ocular lenses varies from 4-15X. The magnification is stamped into the barrel of the lens (Figure 3.2). Total magnification is calculated by multiplying the ocular lens magnification times the objective lens magnification. The objective lenses are on a turret and are closest to your specimen. Magnifications are stamped on the barrel of these lenses. Some common ones are: 4x, 10X, 25X, 40X, and 100X. Most microalgae are observed best at 25X or 40X. The 100X lens usually is an oil immersion lens and a special microscope oil must be used at this magnification. When initially examining the specimen, locate and focus on it at 4X before moving to a higher magnification. This will prevent crushing of the specimen and allow it to be centered in the field of view. Most microscopes are aligned so the move to higher magnification requires little or no focusing, called parfocal. The apparent size of the algae will increase progressively with higher magnification.

It is important to keep lenses clean, especially of salts, which are very corrosive. After examining materials, be sure to gently wipe lenses with lens paper, not Kimwipes or paper towels, which can scratch. Alcohol on a cotton swab wrapped with lens paper may be used to reach dirt inside the lens. In addition, dirt trapped inside the lens may be removed by carefully taking it apart, cleaning, and then reassembling. Care must be taken to correctly place the convex side of the lens in the ocular cylinder. As an additional precaution, keep the microscope in a dry, dust-free place when not in use and cover it with a plastic bag. This will prevent the growth of fungi on the lenses and assure the best operation of your microscope.

Specimen Preparation

Most microalgae are small and numerous, often millions of cells occurring in a drop of water. The easiest technique for examining a specimen is to make a wet mount. A wet mount is prepared by placing a small drop of water on a slide and covering it with a cover slip. The slide should be cleaned prior to use with alcohol (isopropyl or ethanol) and wiped dry. Even new slides should be cleaned, because they can be covered with oil from the manufacturer.

If a large drop is placed on a slide, the cover slip will slide around and the image will be blurry. A thin layer is all that is needed and produces the clearest view. When examining large organisms, which may be crushed by the weight of the cover slip, it is best to prop up one side of the cover slip with a second cover slip. This will prevent crushing, but limits the distance between the objective lens, reducing the magnification attainable. In general, the flatter your preparation, the higher the magnification possible.

Measuring Cell Size

Many descriptions of algae use cell size as a characteristic. Using the microscope and an ocular micrometer, it is easy to determine the length and width of a specimen. An ocular micrometer usually rests on a ridge in the lower part of the ocular, and is held in place by a small threaded ring. To calibrate the ocular micrometer for each magnification, use a stage micrometer for each magnification you use. A micrometer coverslip may also be used as a stage micrometer (available from Florida Aqua Farms). By visually superimposing the stage micrometer markings over the ocular

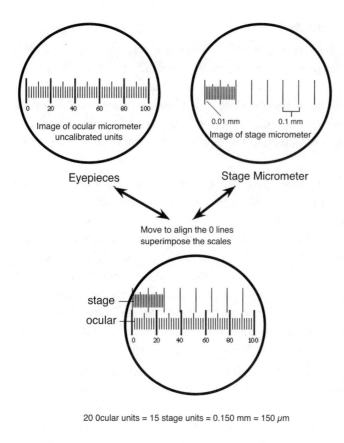

Eyepieces · Stage Micrometer

Move to align the 0 lines
superimpose the scales

20 0cular units = 15 stage units = 0.150 mm = 150 μm

Figure 3.3. Calibration using an ocular micrometer.

micrometer markings, determine the true length between markings in the ocular micrometer, which is in uncalibrated units (Figure 3.3). When measuring a specimen, determine the number of units from the ocular micrometer and then convert to the actual length. The measurements are given in micrometers (μm). One micrometer is one millionth of a meter.

Stains

Motile algae often move too fast for accurate identification and the number of flagella cannot be determined as they swim. To immobilize specimens, especially the flagella, staining is a useful technique. One of the best stains for visualizing and identifying microalgae is Lugol's solution, a mixture of iodine (I_2) and potassium iodide (KI). This mixture of I_2KI and water (5 g I_2 + 10 g KI +100 ml H_2O) is used at dilute concentrations (a light brown color), to preserve algae. This compound also stains starch purple and can be used to distinguish algal groups that store starch versus oil. In addition, many contaminants are difficult to see and staining will allow better visualization of these organisms. Some stains are also available that can be used to determine the viability of algae.

Classification of Microalgae

There are 10 divisions of algae, 9 have unicellular representatives and 6 are commonly used in aquaculture (Figure 3.4). Each division has characteristics that are shared by its members, but its species are distinct enough to be separated from one another. There are 4 major characteristics used to separate the divisions of algae: type of cell wall, presence or absence of flagella, type of photosynthetic storage compound, and the kinds of pigments present (Prescott, 1970; Dawes, 1981; Bold and Wynne, 1985). In addition, the morphology of individual cells and how they attach to form colonies or filaments is important information for distinguishing groups.

Algal cell walls are usually composed of some form of complex polysaccharide, but are a form of glass in other groups, and are absent in some others. Because the composition of an algal cell wall usually requires biochemical determination, this characteristic is not the key factor for aquaculturists in identifying microalgae. Flagella are important characteristics in identifying microalgae. Flagella characteristics are easier to distinguish in motile species. The number of flagella and their point of insertion, either at the end of a cell (apical insertion), or on the side (subapical insertion), are used to distinguish many species. Photosynthetic storage compounds range from starch to lipids, depending on the organism and its culture history, but the precise determination of the type of storage compound requires biochemical techniques that are not practical for most applications.

The most practical characteristic to identify algae is their color or pigmentation (Figure 3.4). These pigments absorb light energy, converting it to biomass by the process of photosynthesis. There are three major classes of pigments and various combinations give each division of algae their characteristic color. The major group of green pigments is the chlorophylls, with chlorophyll *a* as the main pigment that captures blue and red wavelengths of light essential for photosynthesis.

Most carotenoids are protective of the other pigments rather than directly contributing to the photosynthetic reactions. In each division, there are exceptions, such as fucoxanthin in the diatoms and brown algae, which is very active in photosynthesis. The phycobilins are red (phycoerythrin) or blue (phycocyanin) and capture wavelengths that are missed by the other pigments and pass the energy captured onto chlorophyll *a* for photosynthesis.

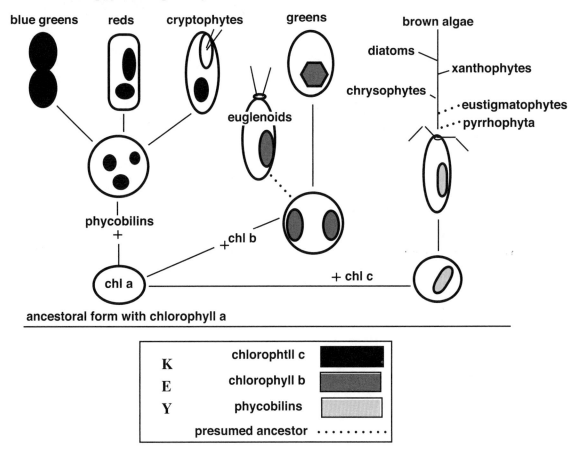

Figure 3.4. The classification of algae basedon pigments.

Several variations in cell shape can be found in algae (Figure 3.5). Unicells can be spherical (Figure 3.5a), compressed laterally (Figure 3.5b), elongated (Figure 3.5c), or box-like (Figure 3.5d) in appearance. In addition, several unicells have arms or spines that are extensions from the cell wall. Many microalgae form filaments of cells connected end to end (Figure 3.5e). Others form ornate colonies of cells that have a distinct pattern and predetermined cell number (a coenobium, Figure 3.5f). Culture conditions will determine the exact morphology of an organism and variation from the typical form is common.

An identification key is provided to the most common alga used in aquaculture. When using this key to identify microalgae, it is best to examine the photomicrograph of each genus with its description. The following alga descriptions are organized into divisions, starting with the simplest forms, the Cyanobacteria and progressing to Chlorophyta, the most abundant group of microalgae. The Chrysophyta is divided into two groups, the diatoms and the golden brown algae, because of differences in their cell walls. Four less common groups, the Rhodophyta, Euglenophyta, Cryptophyta, and Pyrrhophyta also are described.

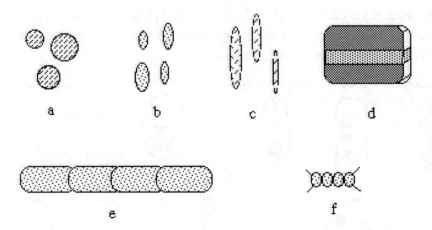

Figure 3.5. Common cell shapes for microalgae.

Key to Genera in the Manual

1a. Cells individual, unicells .... go to number 2

1b. Cells forming a colony or filament .... go to number 14

2a. Cells with flagella...3

2b. Cells without flagella 8

3a. Cells with one flagellum

Euglena (Plate 26)

3b. Cells with two flagella 4

3c. Cells with four flagella

Tetraselmis (Plate 10)

4a. Cells green in color, mostly spherical 7

4b. Cells brown in color, if spherical, with a ridge 5

5a. Cells flatten on one side with one or two flagella

Cryptomonas (Plate 27)

5b. Cells spherical with two flagella

Isochrysis (Plate 23)

5c. Cells with visible cell wall (plates) 6

6a. Cells mostly spherical

Peridinium (Plate 29)

6b. Cells with arms

Ceratium (Plate 28)

7a. Cells mainly in freshwater, spherical

Chlamydomonas (Plate 11)

7b. Cells mainly marine, somewhat elongate

Dunaliella (Plate 13)

8a. Cells red, brown or golden in color...12

8b. Cells green in color...9

9a. Cells spherical...10

9b. Cells lanceolate

Ankistrodesmus (Plate 16)

9c. Cells elliptical

Ellipsoidon (Plate 24b)

9d. Cells cresent shape

Selenastrum (Plate 17)

10a. Cells small, visible U shaped chloroplast 11

10b. Cells large, granular in appearance

Chlorella (Plate 14)

11a. Cells with chlorophyll *a* and *b*

Nannochloris (Plate 12)

11b. Cells with only chlorophyll *a*

Nannochloropsis (Plate 24a)

12a. Cells spherical, reddish in color

Porphyridium (Plate 25)

12b. Cells elongate (naviculoid)

Phaeodactylum (Plate 22)

12c. Cells round or square, depending on view 13

13a. Cells with one pattern on the cell wall

Thallasiosira (Plate 20)

13b. Cells with two patterns on the wall

Cyclotella (Plate 19)

14a. Cells forming a colony of a 4-8 cells (sometimes 2)

Scenedesmus (Plate 15)

14b. Cells forming a filament...15

15a. Filaments golden brown in color..16

15b. Filaments usually blue-green, red or brown 17

16a. Filaments with internal spines connecting cells

Skeletonema (Plate 21)

16b. Filaments with long external spines

Chaetoceros (Plate 18)

17a. Filaments coiled

Spirulina (Plate 7)

17b. Filaments with cells stacked like coins

Oscillatoria (Plate 8)

17c. Filaments with bead-like cells and heterocysts

Anabaena (Plate 9)

Cyanobacteria or Blue-green Algae

Cyanobacteria or blue-green algae are the most primitive group of algae and have both bacterial and algal characteristics. A comprehensive taxonomic description is given by Drouet (1968, 1973). They are prokaryotic organisms, which means they lack the cellular structures (i.e. nucleus, chloroplasts) present in all the other algae. The cell walls of blue-green algae are similar to gram-negative bacteria. In addition, outside the cell wall, several cyanobacteria deposit a protein-polysaccharide capsule that makes them resistant to desiccation and allows them to survive in habitats that are intermittently wet. They possess only chlorophyll *a*, but also have phycobilins as well as various carotenoids. The variety of pigments contributes to the success of cyanobacteria in a variety of habitats having low light and poor water quality. It is the various combinations of these pigments that give the blue-greens a range of colors from green to purple and even red.

Blue-greens never have flagella, but some filamentous forms move by gliding and appear to sway back and forth when in contact with the water surface. Unicellular, colonies, and cyanobacteria filaments are common in aquaculture facilities, both as food and as nuisance organisms. Below are three of the most common.

Spirulina-(Freshwater, Marine; Plate 7) This is a filament 5-6 μm wide and 20-200 μm long in the shape of a coiled spring. It moves by a gliding motion and may be attached or floating, but not free swimming. It may appear blue-green or even red. Several individual cells are present in each coil, but are difficult to distinguish in the filament. *Spirulina* is a component of many flake and pelletized fish and invertebrate foods.

Oscillatoria-(Freshwater, Marine; Plate 8) This blue-green is filamentous, 2-20 μm wide and 10-200 μm long, depending upon the species. It can be straight, bent, curved, or irregularly coiled. It moves by a gliding or swaying motion and may be attached or floating, but not free swimming. It may appear green, blue-green, purple, or red. Individual cells may be difficult to distinguish in the filament, some are longer than wide, others are wider than long. *Oscillatoria* is usually harmless but may cause earthy-musty odors and can be a visible nuisance.

Anabaena-(Freshwater, Marine; Plate 9) This is a filament 3-10 μm wide and 10-200 μm long that is straight, bent, or loosely coiled. It moves by gliding, but this is not common. Cells are typically bead or barrel-shaped. One cell, the heterocyst, is enlarged and has the capacity to fix atmospheric nitrogen. Another enlarged cell, the akinete, is usually found adjacent to the heterocyst and serves as a spore that is resistant to harsh environmental conditions. *Anabaena* is a nuisance organism that is not eaten by most cultured organisms. It may also cause odors.

Chlorophyta-Green Algae

The green algae are the most advanced group of algae and have many characteristics typical of higher plants. They are eukaryotic organisms, meaning they have the typical cellular structures of most algae. All have chloroplasts, their DNA is contained in a nucleus, and some have flagella. The cell wall of green algae is mostly cellulose, although some have no cell wall. They possess chlorophyll *a* and *b* and several carotenoids and are usually grass green in appearance. When cultures become dense and light-limited, cells produce more chlorphyll per cell and become darker green. Most green algae store starch as a food reserve, although a few store oils or other lipids. Unicellular and filaments are the most common forms found in aquaculture facilities. In general, the unicells are food sources and the filaments are nuisance organisms.

Tetraselmis-(Freshwater, Marine; Plate 10) This is a motile green, 9-10 μm wide, 12-14 μm long, with 4 flagella that emerge from a groove at the anterior end of the cell. Cells move rapidly through the water and appear to shake as they swim. It is a 4-lobed cell that is elongate and may have a reddish eyespot. *Pyramimonas* is a closely related green algae that is identical in appearance and swimming characteristics to *Tetraselmis*. Both are popular food sources for culturing rotifers, clams, oysters, and larval shrimp.

Chlamydomonas-(Freshwater, Marine; Plate 11) This is a motile green, 6.5-11 μm wide and 7.5-14 μm long, with 2 flagella that emerge near a swelling or bump at the end of the cell. Cells move rapidly through the water and appear to shake as they swim. Cells are spherical to elongate and usually have a red eye spot. When daughter cells are being formed, the mother cell may lose its flagella and secrete a transparent sack around itself. The mother cell divides and 2-8 daughter cells are formed and contained within the envelope. This stage of the life history may precipitate because it is denser than the culture medium. Daughter cells may be released immediately or this "resting" stage may persist, depending on environmental conditions. Used as a food source for rotifers.

Nannochloris-(Freshwater, Marine; Plate 12) This is a non-motile green, with no flagella. It is a very small, spherical cell, 1.5-2.5 μm in diameter, with few distinguishing features. The chloroplast is usually U-shaped in healthy cells. Cells tend to float in culture, staying in suspension without aeration, which can be an advantage in aquaculture. Popular food source for rotifers, clams, oysters and larval shrimp.

Dunaliella-(Freshwater, Marine; Plate 13) This is a motile green with 2 flagella that emerge near the end of the cell, 5-8 μm wide and 7-12 μm long. Cells move rapidly through the water shaking as they swim. Cells are round to elongate, usually with a red eye spot. The chloroplast appears to occupy about two thirds of the cell, although this may be under environmental control. Reproduction is by simple fission of the mother cell into two daughter cells. Used as a food source for rotifers, clams, oysters and larval shrimp.

Chlorella-(Freshwater, Marine; Plate 14) This a non-motile green with no flagella. It is a medium-sized spherical cell, 2-10 μm in diameter, depending on the species, with a cup-shaped chloroplast. Cells reproduce by forming 2-8 daughter cells in the mother cell which are released depending upon environmental conditions. Although we include several species within the genus *Chlorella*, the only features that link them are cell morphology and reproductive patterns. The species are not necessarily closely related to one another and should be treated as distinct organisms. Several species are used in aquaculture. The most comprehensive study of *Chlorella* was conducted by Kessler (1976). Popular food source for rotifers and daphnids.

Scenedesmus-(Freshwater; Plate 15) This is a non-motile green that is usually a 4-celled, flattened colony, 12-14 μm wide and 15-20 μm long. Cells are elliptical to lanceolate (long and slender), some with spines or horns. Each cell produces a new 4-celled colony when it reproduces. It is often a nuisance organism and grows well in nutrient-rich water. It is not usually cultured as a food source.

Ankistrodesmus-(Freshwater; Plate 16) This is a non-motile green that is usually single-celled with long, thin, crescent shaped cells. It also occurs in colonies of 4 to 8 cells arranged at right angles to one another. This organism is a common contaminant of water supplies and can live in water pipes, water jugs, and stock solutions. It is not usually cultured as a food source.

Selenastrum-(Freshwater; Plate 17) This is a non-motile green, 2-4 μm wide and 8-24 μm long. It is a crescent-shaped single cell that may also be twisted. This organism has been extensively used for algal bioassays in aquatic toxicology because it is easy to grow. It requires only inorganic nutrients and does not form colonies (Palmer, 1977). This organism is occasionally used as a food source for daphnids.

Chrysophyta-Diatoms

Diatoms are a unique group of algae with a cell wall composed of silicon dioxide, similar to the mineral opal in composition. The cell wall is perforated with numerous holes that make these organisms ideal sieves and they are used commercially in several filtering devices. Two major groups are based on the symmetry of the wall, either bilateral or radial (Patrick and Reimer, 1966). They have typical higher plant-like characteristics, including eukaryotic cellular organization. No flagella are found except in the reproductive cells of certain species. In the bilaterally symmetric forms, some move by gliding when in contact with a substratum. All have chloroplasts and their DNA is contained in a nucleus. They possess only chlorophyll a and c, as well as several carotenoids such as fucoxanthin which give them a brownish color. Several store lipids and oils which are nutritionally beneficial to cultured animals. Unicells and filaments may be encountered in aquaculture facilities as food sources. Brown patches on the sides of culture vessels are usually diatoms and can be removed with a razor, scouring pad, or acidic solutions.

Chaetoceros-(Marine; Plate 18) This is single-celled and may form chains by interconnecting spines of adjacent cells. The main body of the organism is shaped like a petri dish. When viewed from the side, individual cells appear square, 12-14 μm long and 15-17 μm wide, with spines protruding from the corners. Cells may form chains of 10-20 cells may reach 200 μm in length. When cultured with strong aeration, *Chaetoceros* does not form colonies. Large cultures are brown in appearance and individual cells are golden-brown in color, surrounded by a translucent cell wall. Popular food source for rotifers, clams, oysters and larval shrimp.

Cyclotella-(Freshwater, Marine; Plate 19) This is a radially symmetric unicellular form, 5-12 μm in diameter which rarely forms chains. It is tire-shaped and has distinct markings on the surface and edge of the cell. Spines are rare and usually not visible with the light microscope. Both freshwater and marine forms are common. Several species grow well in enriched cultures, and are occasionally used as a food source.

Thallasiosira-(Marine; Plate 20) This is a radially symmetric form, 11-14 μm wide and 14-17 μm long, that usually exists as unicells but may form chains. In chains, several cells are stacked together and these stacks are connected by a thin thread-like structure to make a filament. A common planktonic form and aquaculture food organism. Used as a food source for clams, oysters, and larval shrimp.

Skeletonema-(Marine; Plate 21) This a chain-forming organism that has roundish cells connected by long silica strands. Individual cells are 6-10 μm wide and 20-25 μm long, with filaments ranging to 500 μm long with 15-20 cells. This organism is also found in brackish water with salinity as low as 10 ppt and is a common planktonic genus and aquaculture food organism. Used as a food source for clams, oysters, and larval shrimp.

Phaeodactylum-(Marine; Plate 22) This diatom is bilaterally symmetrical, and has two forms; one that is boat-shaped, 2.5 to 5 μm wide and 12-25 μm long, and another that is somewhat triangular. The bilateral form is termed naviculoid and many non-food diatoms have this general appearance. These forms are motile when in contact with a substratum. They are occasionally used for rotifers, clams, and oysters and are generally common contaminates.

Chrysophyta-Golden-Brown Algae

The Golden-Brown algae are related to the diatoms but lack a silica cell wall during most of their lives. They have the typical characteristics which are found in most of the algae. Some members of this group have flagella and are motile (Bourelly, 1968). All have chloroplasts and their DNA is contained in a nucleus. They possess only chlorophyll a and c and several carotenoids such as fucoxanthin which give them a brownish color. Several store lipids and oils that are nutritionally beneficial to cultured animals. Unicellular forms are commonly cultured in many aquaculture facilities as food sources.

Isochrysis-(Marine; Plate 23) This is a motile cell with 2 flagella that emerge near the end of the cell. Cells move rapidly through the water, rotating as they swim. It is round organism, 4-8 μm in diameter, golden in color and usually has a red eye spot. The chloroplast is cup-shaped and appears to occupy about one third of the cell, the rest appears clear. Reproduction is by simple fission of the mother cell into two daughter cells. Popular food source for rotifers, clams, oysters, and larval shrimp.

Nannochloropsis-(Freshwater, Marine; Plate 24a) This is a nonmotile greenish colored cell with no flagella. It is a small, spherical cell, 4-6 μm in diameter, with few distinguishing features. The chloroplast usually occupies much of the cell. Cells tend to float in culture and stay in suspension without aeration. This organism is placed in a separate division from *Nannochloris* because it lacks chlorophyll *b*. Even though they have similar names, *Nannochloropsis* and *Nannochloris* are as different as fish and shrimp are from one another. Popular food source for rotifers, *Artemia*, and filter feeders in general.

Ellipsoidon-(Freshwater, Marine; Plate 24b, insert) This is a nonmotile greenish colored cell with no flagella. It is a small, elliptical cell, 4-6 μm in length, with one distinct chloroplast, which occupies half of the cell. Cells tend to float in culture and stay in suspension without aeration. This organism is placed the same group as *Nannochloropsis* because it also lacks chlorophyll *b*. This organism stores oil and may compliment the nutritional needs of cultured organisms. Used as a food source for clams and oysters.

Rhodophyta-Red Algae

The Rhodophyta or red algae are common macroalgae, but there are only a few microscopic species. They possess only chlorophyll *a* but also have other pigments like phycocyanin, (a blue pigment), and phycoerythrin (a red pigment), as well as various carotenoids. Phycoerythrin gives red algae their appearance, although some are bluish-green to purple. Unicellular red algae have no flagella and are not motile. Unicells may be used in aquaculture facilities.

Porphyridium-(Marine; Plate 25) This a unicellular form that is spherical, 7-12 μm in diameter. It is classified as one of the simple species of red algae because is does not reproduce sexually and has glycogen as a storage compound. As with all reds, it is not motile. It appears reddish or blue-green in culture. It has been used in aquaculture for the production of carbohydrates.

Euglenophyta

The Euglenophyta are placed in the green algae by some taxonomists and in the protozoans by others because they have both plant and animal characteristics. They are eukaryotic organisms, with the typical cellular structures found in most algae, but they also have a gullet which allows them to ingest particles. Most Euglenoids can survive in darkness by feeding on bacteria or small organic particles. They have one long flagellum, and they usually swim by pulling themselves through the water. In addition, some have a distinct amoeboid movement. Euglenoids have no cell wall, but have a rigid protein outer layer, the pellicle, that serves the same purpose as the cell wall. The pellicle is visible as "strips" that traverse the cell. They possess chlorophyll *a* and *b* , several carotenoids and are usually grass green in appearance. Euglenoids store the carbohydrate paramylum as food reserve and a few have oils and other lipids. Only unicells are present in aquaculture facilities, both as food sources and as nuisance organisms. *Euglena* is common in nutrient rich habitats.

Euglena-(Freshwater, Marine; Plate 26) This is a motile green with 1 flagellum that emerges from a gullet near the end of the cell. Cells move slowly through the water and appear to twist or rotate as they swim. Most species are elongate, 10-15 μm wide and 50-150 μm long, and usually have a red eye spot which is evident. The storage compound appears as shiny elliptical bodies in the cell. The chloroplasts are numerous and discoid. Generally, not used as a food source.

Cryptophyta

The Cryptophyta algae are a unique group of unicellular forms that are apparently not closely related to other groups of algae (Butcher 1967). They are eukaryotic organisms, but they also have a gullet allowing them to ingest food. The gullet is lined with ejectosomes that may be used to immobilize small organisms such as bacteria. Members of this group have flagella and are motile. All have one to two chloroplasts and they possess chlorophyll a and c, phycocyanin and phycoerythrin, as well as several carotenoids that give them a brownish color. Several store lipids and oils that are nutritionally beneficial to cultured organisms. Cells may be utilized in aquaculture facilities as food sources.

Cryptomonas-(Freshwater, Marine; Plate 27) This genus is the most common cryptomonad, having one long and one short flagellum emerging off-centered from a flattened area of the cell. This flattened area is the most distinguishing feature of this group of organisms. Cells range in size from 8-16 μm long to 6-8 μm wide. Cells swim slowly in large sweeping loops through the water. The have 1-2 brown chloroplasts and can photosynthesize or survive on bacteria. Generally not used as a cultured food source, but natural populations are consumed by rotifers, clams, oysters, and larval shrimp.

Pyrrhophyta

The Pyrrhopyta contains the dinoflagellates, a unique group of unicellular organisms that have two flagella and are common in both freshwater and marine habitats. They are eukaryotic organisms, with members that are photosynthetic, saprophytic, parasitic, and symbiotic. Most members of this group are motile, although nonmotile forms are often part of the life cycle in many species. Photosynthetic pigments in this division include chlorophyll a and c, the xanthophylls peridinin and dinoxanthin, as well as several others. Most species store starch as a food reserve. One of the unique features of this group of organisms is the cell wall, which is made of cellulose plates that give them an armored appearance, although some forms have no cell wall. A pronounced transverse groove encircles the cell and houses one of the two flagella and a second flagella trails behind the organism. Movement is usually smooth in one direction or with some rotation. Many organisms also have trichocysts, proteinaceous structures that can be discharged from the surface of the cell for protection from predation. The phenomenon "red tide" is attributed to blooms of dinoflagellates because of the reddish pigments that accumulate in these organisms and the high numbers that occur under certain environmental conditions. Some dinoflagellates also cause paralytic shellfish poisoning and accumulate high concentrations of neurotoxins. Some species are fish parasites, causing problems such as "velvet disease." Most species are not food organisms because they are too large to be consumed, however, they are common contaminates in marine systems.

Ceratium-(Freshwater, Marine Plate 28) This genus is one of the most common and easily recognizable dinoflagellates, heavily armored with one long arm and two shorter ones that are straight or curved, depending on the species. Cells are motile and swim actively when first examined, usually in a straight or curved direction. Cells range in size from 30 to 90 μm long and 10 to 30 μm wide. This organism is a common contaminant in marine cultures.

Peridinium-(Freshwater, Marine; Plate 29) This genus is a common dinoflagellate, heavily armored, mostly round in appearance, with a small apical crest and spine-like legs in some species. A distinct transverse groove is usually easy to see, which is characteristic of this organism. Cells are motile and swim actively in a whirling motion. Cells range from 25 to 80 μm in diameter. This organism is a common contaminant in marine cultures.

Contaminants

Several contaminants occur in microalgae cultures. Some are harmless, while others may cause serious problems for the microalgae or the animals being cultured. Several precautions to minimize contamination should be followed when growing microalgae.

Protozoans

There are two major groups of protozoans that are common contaminants in aquaculture: ciliates and amoeboid forms. Most are solitary, but some are colonial. Free-living forms are the most common in aquatic systems and some parasites are also found. Some of these merely compete for nutrients, whereas others eat microalgae or accumulate toxins that can kill rotifers, *Artemia*, and fish. The protozoans form a food vacuole that digests ingested food internally. Cilia, flagella, and pseudopodia are used for locomotion. For more information on identification consult Lee, et al. (1985).

Class Ciliata

The ciliates utilize cilia, small hairlike structures, for locomotion and sensing their surroundings. They are numerous and very diverse in both freshwater and marine environments. Several species may be encountered in aquaculture.

Order Holotricha

Paramecium-(Freshwater, Marine; Plate 30) In general, the cells are cigar-shaped, 25-40 μm wide and 50-150 μm long, and have a distinct oral groove. Freshwater forms have a contractile vacuole that helps osmoregulate and excrete nitrogenous compounds. Many eat algae, bacteria, and smaller ciliates. Some species are foot-shaped and several can form resting cysts.

Colpoda-(Freshwater, Marine; Plate 31) The cells in this genus are kidney-shaped, 15-20 μm wide and 25-30 μm wide, and appear to have a flattened end. They are active swimmers and move through the water in sweeping arcs. They actively eat microalgae and are common contaminants in aquaculture.

Order Spirotricha

Euplotes-(Freshwater, Marine; Plate 32) This genus is placed in the Spirotricha, and has cilia modified into special structures called cirri. The cirri are used in locomotion, as well as to sense the environment. Two forms of movement are observed; they crawl slowly on a surface and swim rapidly in the water. This genus is common in aquaculture and the large cells, which range from 40-50 μm wide and 90-120 μm long, are sometimes eaten by larval fish.

Order Sarcodina

The sarcodina include the protozoans that use pseudopods for locomotion and feeding. They engulf other protozoans, algae, and even rotifers as well as dead organic matter. This group includes the Foraminifera which secrete a shell made of calcium and the Radiolaria which have a shell made of silicon.

Order Amoebina

Amoeba-(Freshwater, Marine; Plate 33)   This genus crawls on various substrata and engulfs any organic matter that is small enough to be surrounded by its pseudopodia, arm-like extensions of its body. Cells vary in size from 20-70 μm in length. Although they are usually not numerous in aquaculture, they are persistent contaminants.

Bacteria (Plate 34)

Bacteria are found in freshwater and marine environments and are important in many biological processes (Brock, et al. 1984). They are related to the cyanobacteria or blue-green algae, however they do not contain chlorophyll *a* and are generally not photosynthetic. One group of bacteria degrades organic matter, producing a characteristic rotten egg smell and turning surfaces black. They are common inhabitants of microalgae cultures and may even contribute to the success of these cultures by leaking essential vitamins into the medium. The number of bacteria in a microalgae culture is usually small during exponential growth and increases as algal cells die and release organic compounds to the medium. Many bacteria can co-exist on the excreted substances from the algae.   Occasionally bacteria may dominate a culture before the algae become established and these cultures appear white and cloudy. Some organisms, like protozoans and some rotifers and Cladocera feed on bacteria, but other animals find them too small at 0.5-3 μm in length.

Nematodes (Plate 35)

Occasionally, small roundworms are found in microalgae cultures, but generally do not cause a problem. Nematodes are usually parasitic, however, free-living forms are found in the aquatic environments that actively eat bacteria and microalgae. Some nematodes (e.g. microworms) are good food for fish.

Fungi

Fungi include the molds, mildews and mushrooms. In older cultures as well as cultures enriched with glucose or other sugars, fungi are common inhabitants and may cause problems. Little is known about this group and its effects on algal cultures.

Rotifers (Plate 36)

Rotifers are important food sources for larval fish and are often cultured adjacent to microalgae cultures. Microalgae are excellent food for rotifers, however, rotifers may also be considered contaminants if they appear in microalgae cultures at the wrong time. Even a small inoculum of rotifers can destroy an algal culture in a short time.

Photomicrographs

The following photomicrographs were taken with a Leitz microscope magnified between 250 and 1000 times. On each of the photographs, the bar is 10 μm long. Lugol's was used to stain some of the cells, especially the motile forms. Complete descriptions of each organism can be found in Classification of Algae and Contaminants section.

Organism	Photo	Group
Spirulina	7	Blue-green
Oscillatoria	8	Blue-green
Anabaena	9	Blue-green
Tetraselmis	10	Green
Chlamydomonas	11	Green
Nannochloris	12	Green
Dunaliella	13	Green
Chlorella	14	Green
Scenedesmus	15	Green
Ankistrodesmus	16	Green
Selenastrum	17	Green
Chaetoceros	18	Diatom
Cyclotella	19	Diatom
Thallasiosira	20	Diatom
Skeletonema	21	Diatom

Organism	Photo	Group
Phaeodactylum	22	Diatom
Isochrysis	23	Golden Brown
Nannochloropsis	24	Golden Brown
Porphyridium	25	Red
Euglena	26	Euglenoid
Cryptomonas	27	Cryptomonad
Ceratium	28	Dinoflagellate
Peridinium	29	Dinoflagellate
Paramecium	30	Protozoan
Colpoda	31	Protozoan
Euplotes	32	Protozoan
Amoeba	33	Protozoan
Bacteria	34	Monera
Nematode	35	Nematode
Rotifer	36	Rotifera

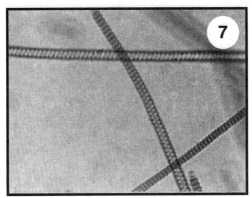

Spirulina

Spirulina - Blue green

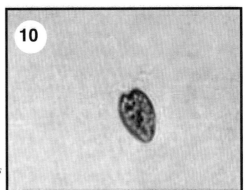

Oscillatoria - Blue green

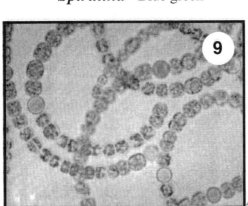

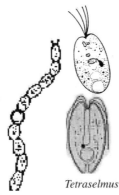

Anabaena

Tetraselmus

Anabaena - Blue green

Tetraselmis- Green

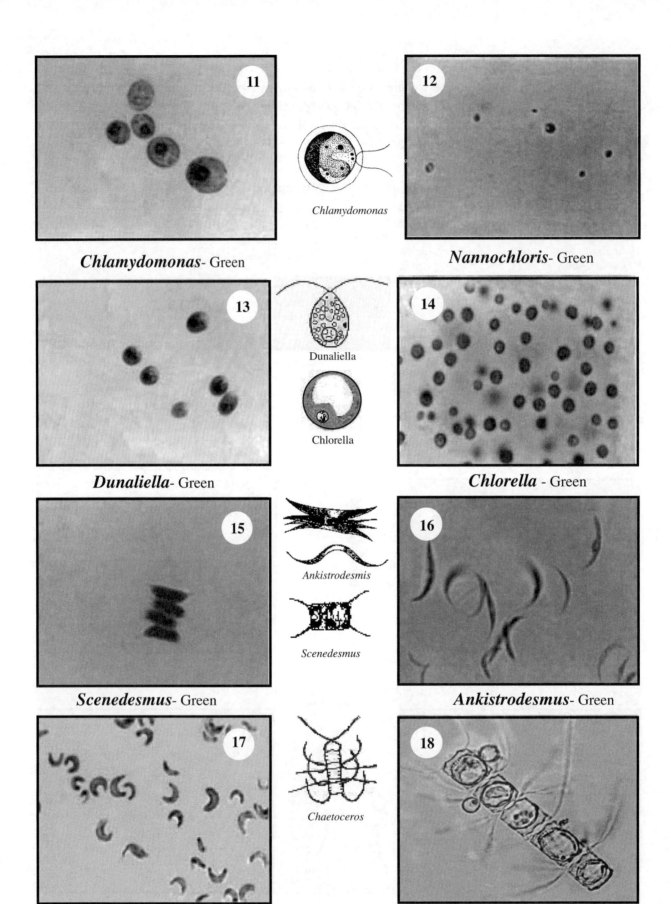

Chlamydomonas- Green

Chlamydomonas

Nannochloris- Green

Dunaliella- Green

Dunaliella

Chlorella

Chlorella - Green

Scenedesmus- Green

Ankistrodesmis

Scenedesmus

Ankistrodesmus- Green

Selenastrum - Green

Chaetoceros

Chaetoceros - Diatom

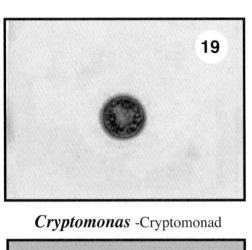

19

Cryptomonas

Cryptomonas -Cryptomonad

20

Thallasiosira - Diatom

21

Skeletonema

Phaeodactylum

Skeletonema - Diatom

22

Phaeodactylum - Diatom

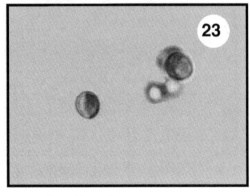

23

Isochrysis

Isochrysis - Golden Brown

24 a

24b

Ellipsoidon

Nannochloropsis - Golden Brown

25

Euglena

Porphyridium - Red

26

Euglena - Euglenoid

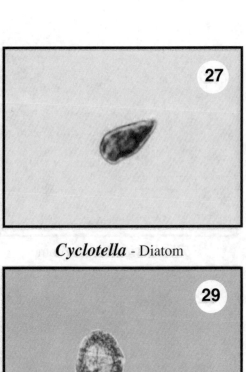

Cyclotella - Diatom

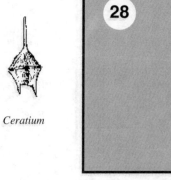

Ceratium

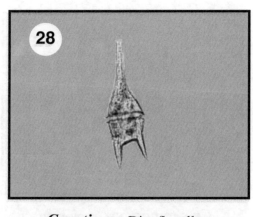

Ceratium - Dinoflagellate

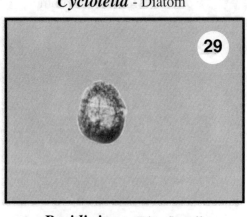

Peridinium

Peridinium - Dinoflagellate

Paramecium

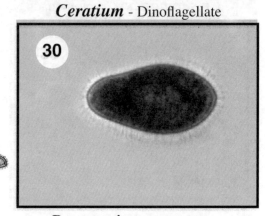

Paramecium - Protozoan

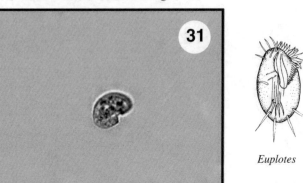

Euplotes

Colpoda - Protozoan

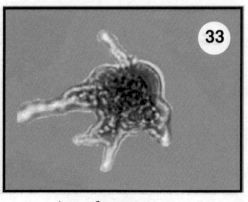

Amoeba

Amoeba - Protozoan

Euplotes - Protozoan

Bacteria - Monera

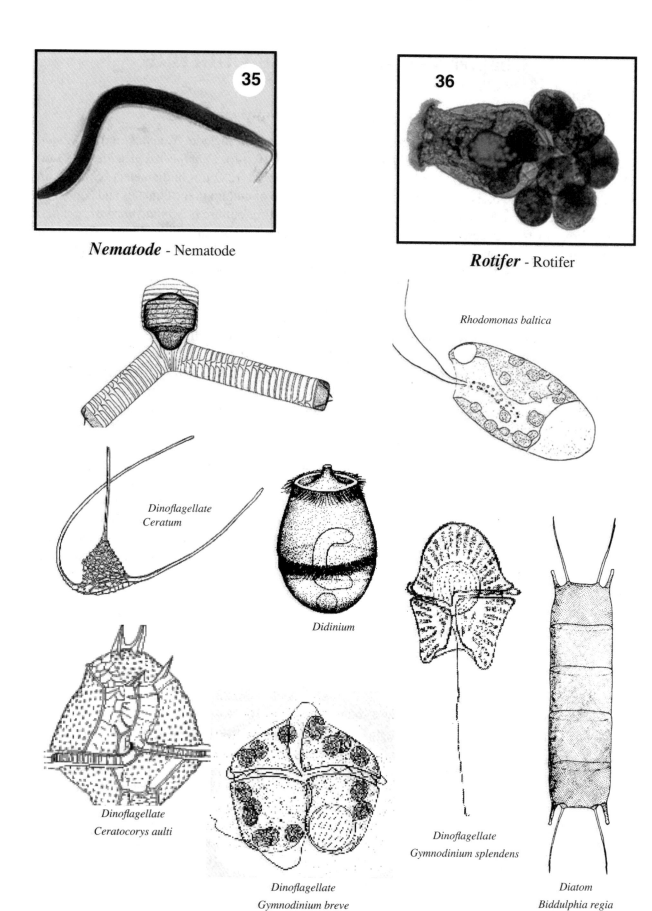

35

Nematode - Nematode

36

Rotifer - Rotifer

Dinoflagellate
Ceratum

Didinium

Rhodomonas baltica

Dinoflagellate
Ceratocorys aulti

Dinoflagellate
Gymnodinium breve

Dinoflagellate
Gymnodinium splendens

Diatom
Biddulphia regia

Chapter 4 - MICROALGAE CULTURE

Introduction

Utilization in aquaculture

Currently, the most popular algae species utilized in aquaculture include: *Nannochloropsis oculata* (2-4 μm), *Isochrysis galbana* (5-7 μm), *Tetraselmis chuii* (7-10 μm), *Chaetoceros gracilis* (6-8 μm), *Dunaliella tertiolecta* (7-9 μm), and several species of *Chlorella* (3-9 μm in diameter). Species are usually selected on the basis of size, nutritional value, and ease of culture in the climates and conditions where they are used. The most widely used microalgae for rotifer culture is *Nannochloropsis oculata*, often referred to as "Japanese *Chlorella*" or "marine *Chlorella*" (Maruyama et al. 1986). This confusion of names is unfortunate because *Nannochloropsis* is not very closely related to *Chlorella* phylogenetically. Ecologically, *Chlorella* is actually considered a soil or attached alga rather than a planktonic (free floating) form, but it can be easily maintained in suspension using heavy aeration.

Nutrient Uptake

Most species of algae are photoautotrophs that derive energy from light and carbon dioxide to obtain carbon atoms. Others are true heterotrophs that do not require light and can acquire energy and carbon from organic compounds such as sugars and organic acids. Yet, many other species are mixotrophic and can reproduce in light or darkness. Heterotrophic and mixotrophic microalgae are found in virtually every taxonomic class. Photoautotrophic algae that also have ability to obtain energy heterotrophically often grow at increased rates by simultaneously utilizing light and carbon dioxide as well as organic substances (Droop 1974). Under commercial pharmaceutical applications densities 1000 times higher can be achieved then under phototrophic conditions. Some algae have been labeled as "obligate photoautotrophs" based on experimentation at ambient nutrient levels. Even some of these species can become heterotrophic when nutrient levels are substantially above or below natural conditions (Gladue, 1991). Examples are *Brachiomonas submarina* and some strains of *Haematococcus pluvialis* which can reduce nitrate when growing photosynthetically, but are unable to do so in the dark. When supplied with a reduced nitrogen source such as ammonia, they were found to grow heterotrophically (Neilson and Lewin 1974). Other examples are *Prymnesium parvum* and *Chroomonas (Pyrenomonas) salina* which are unable to grow heterotrophically on low natural levels of glycerol, but can do so when supplied with very high concentrations (0.25 M) (Droop 1974). Many microalgae are grown heterotrophically to produce pharmaceutical products.

Algal growth media

Chemical features of the aquatic environment play an important role in determining algae growth rates and quality. Microalgae and macroalgae derive nutrients from the surrounding water. They have no roots like land plants, but are able to absorb nutrients directly across their cell membranes.

Since the objective of culturing microalgae is to obtain the highest density in the shortest possible time, the utilization of natural seawater or freshwater with limited enrichment will not yield good results. In natural waters, trace metal concentrations are usually sufficient, but either nitrogen ans/or phosphorus is usually limiting. Phosphorous limitation is more common in freshwater, whereas nitrogen limitation is more common in the sea. In some waters, silica can be in low enough concentration to limit diatom growth (Darley 1982). Cultured microalgae grow best in media with quite different primary and trace nutrient composition than natural waters. Nitrate levels in intensive microalgae cultures are commonly 100 to 1000 times higher than that found in nature. However, enriched natural seawater often produces higher densities than artificial saltwater or deionized freshwater solutions. Indicating that undefined dissolved nutrients in natural waters may enhance growth

Several nutrient enrichment media for microalgae have been described, many of which are based on natural seawater enriched with soil extract, nitrates and phosphates, and specific metals (Stein 1973). In recent years, Guillard's F/2, a four-part enrichment medium, consisting of inorganic nitrates, phosphates, trace metals, and vitamins, has become popular in aquaculture (Table 4.5). Over the years, there have been several composition modifications to this formulation, including a one-part liquid and wet/dry formulation (Florida Aqua Farms), and a two-part liquid formulation (Fritz Chemicals and Kent Marine) These formulations provide most microalgae all the nutrients required for rapid growth, similar to the conditions found in eutrophic lakes or nutrient-enriched seawater.

Microalgae growing in intensive cultures are therefore quite different from those growing in nutrient-limited natural waters. Natural algae populations are subject to marked changes in temperature, light intensity and duration, leading to unpredictable growth rates. Its is usually possible for aquaculturists to improve both predictability of microalgae growth and biomass yield by using controlled intensive culture methods. Higher cell densities and more consistent nutritional quality are frequently the results of intensive culture. Generally, smaller culture vessels offer more control over the culture environment and algae yields. Scaling up from small flasks to large tanks or even outdoor ponds cannot be done without culture modifications, sacrifice of control, loss of nutrient balance, and yield loss per unit volume. Outdoor cultures are usually supplied with only the barest essentials, and agricultural-grade rather than laboratory or technical grade fertilizers are often used (Gonzalez-Rodriguez and Maestrini 1984). Algae densities are usually lower in larger culture vessels, but what is most important is the number of alga cells per unit volume. For example, in a 100-liter vessel you consistently reach 20 million cells/ml, but in a 1000-liter vessel you may only be able to achieve 5 million cells/ml in the same period of time. Based on the fact that smaller vessels are easier to control, it is usually better to use five 100-liter culture vessels rather than one 500-liter vessel. Scaling up presents new problems which need further research.

Pigments

The primary pigments of green microalgae which are involved in photosynthesis are the chlorophylls. Chlorophylls respond to the blue and red wavelengths of visible light, with absorption peaks at about 460 and 665 nm. Different types of algae may have other pigments in addition to the chlorophylls, like xanthophylls, carotenes, and phycobilins. These pigments absorb light in the yellow and orange range then transfer this energy to chlorophyll. Effectiveness of photosynthesis to produce compounds necessary for microalgae growth is determined by light intensity, spectral quality, and nutrient availability.

Algae Species	EPA	Total Omega 3
Nannochloropsis oculata	30.5%	42.7%
Pavlora lutheir	13.8%	23.5%
Skeletonema costratum	13.8%	15.5%
Phaeodactylum tricornutum	8.6%	9.6%
Tetraselmus tetratheie	6.4%	8.1%
Isochysis galbana	3.5%	22.5%
Isochrysis aff galbana	0.5%	3.3%

Table 4.1. Fatty acid composition of some species of phytoplankton used as feed in Japan. Numbers shown as % total fatty acids.

Nutritional Value of Microalgae

Microalgae composition

In general, the nutritional value of microalgae is directly related to the species, nutrient supply, light, and other physical and chemical conditions during growth. For example, when *Monodus subterraneus* is growing exponentially, it has high photosynthetic and respiratory rates, a protein content of almost 70% (dry weight), high levels of chlorophyll and nucleic acids, but low levels of carbohydrates and fats (Fogg 1959). In contrast, nitrogen-deficient conditions produce *Monodus* cells which have low photosynthetic and respiratory rates, are less than 10% protein, have low levels of chlorophyll and nucleic acids

Table 4.1A - Nutrient Analysis of Six Microalgae

SPECIE	Nannochloropsis	Tetraselmus	Pavola	Isochrysis	Thalassiosira	Cheatocerus
Dry Weight	18.4%	18.9%	9%	9%	9%	4.5%
Calories/10ml	44.4	48.2	45+/-	45.5	22+/-	16.2
Protein	52.11%	54,66%	51.61%	49.69%	50%+/-	27.68%
Protein*	??	30-26%	58-62%	41-47%	??	35-38%
Carbohydrate	16%	18.31%	23%+/-	24.15%	??	23.2%
Carbohydrate*	??	27.06%	15.31%	22.54%	??	19.40%
Ash*	??	29.6%	7.4%	8.4%	??	14.7%
LIPIDS	---	---	---	---	---	---
Lipids total	27.64%	14.27%	19.56%	17.07%	??	9.29%
Lipids total*	??	5.12%	15.21%	22.54%	??	??
Omega 3	42.7%	8.1%	23.5%	22.5%	??	??
EPA C20:5n3	30.5%	6.4%	13.8%	3.5%	??	??
EPA C20:5n3*	31.42%	9.3%	21.00%	2.50%	??	??
DHA C22:6n3*	0%	0%	8.20%	10.20%	??	??
ARA C18:2n6*	5.26%	0.40%	??	0.52%	??	??
VITAMINS	---	---	---	---	---	---
Vitamin C*	0.85%	0.25%	??	0.4%	??	1.60%
Vitamin C	0.90%	??	??	0.98%	??	1.62%
Chlorophyll A*	0.89%	1.42%	??	0.98%	??	1.04%
AMINO ACIDS	---	---	---	---	---	---
Aspartic*	9.4%	8.95%	7.27%	10.36%	12.18%	??
Serine*	4.32%	3.71%	3.33%	3.86%	7.42%	??
Glutamic*	15.48%	17.59%	15.47%	11.42%	17.13%	??
Glycine*	7.11%	5.93%	4.75%	6.78%	7.83%	??
Histidine*	0.61%	0.19%	0.50%	0.23%	0.43%	??
Arginine*	4.57%	4.59%	4.11%	4.39%	2.83%	??
Threonine*	5.28%	4.40%	3.51%	4.88%	5.38%	??
Alanine*	1.54%	7.62%	5.66%	7.55%	8.57%	??
Proline*	15.12%	6.47%	7.34%	12.48%	8.65%	??
Tyosine	1.06%	1.84%	1.29%	2.84%	0.60%	??
Valine*	6.90%	5.00%	4.56%	5.49%	5.02%	??
Methionine*	2.64%	2.55%	2.10%	3.79%	2.01%	??
Lysine*	9.07%	8.20%	4.65%	6.96%	7.59%	??
Isolucine*	1.47%	0.97%	0.16%	1.11%	1.78%	??
Leucine*	11.57%	7.94%	7.19%	6.87%	9.21%	??
Phenylalanine*	1.92%	3.51%	26.72%	8.25%	1.25%	??
Taurine*	0%	10.55%	0%	0%	0%	??

Special Note: algae profiles change due to varying nutrients, sunlight, temperature, and photoperiod. Information with (*) from Reed Aquaculture on algae grown on a modified Guillards f/2 formula.

Algal Species	Protein Nitrogen x 6.25	Fat	Carbohydrates	Ash
Chaetoceros muelleri	34.75 - 38.50	33.15	19.40	14.7
Dicrateria sp.	38.06	29.09	22.45	10.4
Isochysis galbana 3011	41.53 - 46.81	22.54	22.54	8.4
Pavlova viridis	58.51 - 62.25	15.31	15.04	7.4
Tetraselmis sp.	30.06	5.16	26.68	38.1
Tetraselmis subcordiformis	46.38	5.09	27.43	21.1

Table 4.2. Analysis of nutrient components of six species of microalgae (% dry weight of cells).

and high levels of carbohydrates and fats (Fogg 1966). Rapidly growing *Botryococcus braunii* cultures are bright blue-green in nitrogen-sufficient conditions, but actually change to a brick orange when nitrogen starved. This color change occurs as other microalgae pigments like carotenoids mask the green of chlorophyll. Therefore, it is important to feed microalgae to zooplankton before their nutrient supply is exhausted.

Different microalgae cultured under the same conditions differ in fatty acid composition. Table 4.1 (Okauchi 1991) shows the fatty acid composition of several of the more popular microalgae used in aquaculture. *Nannochloropsis oculata* or *N. salina* are currently considered important algae species because of their high levels of B_{12} and eicosapentaenoic acid (EPA). It is especially important to note the difference between the two *Isochrysis galbana* strains listed. *Isochrysis galbana* was selected by farmers in Japan because of vigorous growth rate in hot weather, but its EPA content is substantially different from another strain, *Isochrysis* aff. *galbana*. Brown 1991 showed a total lipid content of 7.0 for *Isochrysis galbana* and 5.9 for *Isochrysis* aff. *galbana* but does not present the EPA and HUFA levels of the total lipid content.

Chen (1991) provided a summary (Table 4.2) of nutritional constituents for several microalgae used in aquaculture in the Republic of China. Again there are significant differences between various genera and even species within the same genus. Brown and Jeffrey (1992) provide information on the biochemical composition of several chlorophytes (*Chlorella, Stichococcus*) and prasinophytes (*Tetraselmis, Pyramimonas*). Dunstan, et al. (1992) list lipid and fatty acid analysis for many of these same species. The biochemical composition of *Isochysis galbana* has been described (Sukenik and Wahnon 1991), as well as *Nannochloropsis sp.* (Sukenik, et al. 1993). Brown et al. (1997) described biochemical composition of about 40 species of microalgae from seven algal classes.

Herrero et. al. (1991) examined the variation of 4 ediums (Waine, ES, F/2, Algal-1™) and for *I. galbana* the varience of cell density was 10.11 to 16.15, protein 5.17 to 9.57, carbohydrates 4.28 to 5.59 and lipids from 20.68 to 28.38.

Microalgae are a significant source of vitamin C (0.11-1.62% dry weight). The diatom *Chaetoceros gracills* contains about 45% more than *Nannochloropsis oculata* (1.62 to 0.9%) while *Isochysis* (TISO) is about half that of *Nannochloropsis* (0.45%).

Physical Requirements

Environmental variations

Microalgae are hardy plants, yet still require specific environmental conditions to grow well. As with any living organism, physical conditions have a large influence on the growth of microalgae. Each species has a particular range of temperatures, light intensities, spectral preferences, salinities, and oxygen/carbon dioxide ratios that yield maximum growth. It takes a great deal of controlled experimentation to determine the optimal growth conditions for a particular species, but fortunately, this information is known for many species valuable to aquarists and aquaculturists.

Physical requirements vary greatly. For example, some microalgae grow at temperatures below 10°C (50°F) while some blue-greens grow best around 60°C (140°F). Optimal growth temperatures vary with alga species, light intensity, nutrient concentration and composition, as well as acclimation temperature (Fogg 1966). Some green and blue-green algae in certain media will grow in total darkness and some grow in light in excess of 10,000 lux (10.764 lux = 1 lumen/sq ft = 1 foot-candle). Continuous agitation may be required for species found living in swift currents. Other forms live in seafoam at salinities as high as 85 ppt (parts of salt per thousand parts of water).

Circulation patterns

Circulation patterns in a culture vessel may play an important role in microalgae production. Fogg & Than-Tun (1960) found better growth of the blue-green algae *Anabaena cylindrica* shaken at 90 rather than 65 oscillations per minute, but growth stopped at 140 oscillations per minute. Subtle changes in circulation can therefore have large effects on yield. Our research indicates better growth is often achieved when the air stone is removed, allowing large bubbles pass through the water column. Small bubbles create foam or fine spray which trap algae cells and deposit them on the sides of culture vessels, causing desiccation and death.

Light intensity

A light intensity of 2500-5000 lux (250-500 footcandles) is optimal for microalgae. Light output from a bulb is often listed in lumens which is synonymous with the term lux. Conversion of foot-candles to lux or lumens is: one ft. c = 10.8 lumens or lux. Output of a standard new "Cool-White" fluorescent bulb is about 2500 lux or 232 foot candles. Yet, the spectral qualities or wavelengths of the common "cool white" bulb is poor for intense microalgae culture. Most algae respond to red and blue light. Brown colored algae actually respond better to blue wave lengths. Bulb aging is very important since its output and spectral proportions change over time. At least yearly bulb replacement is essential.

Light shading by algae cells may become limiting as density increases. At high densities, incoming light is shaded from all but the cells currently at the outer surface of the container. Shaded cells become light-limited as their photosynthetic capacity is diminished. There-

Algal Species	Temp °C	Light Lux	Salinity ppt
Chaeotoceros muelleri	25°-35°	8,000-10,000	20 - 35
Phaeodactylum tricornutum	18°-22°	3,000-5,000	25 - 32
Dicrateria sp.	25°-32°	3,000-10,000	15 - 30
Isochysis galbana	25°-30°	2,500-10.000	10 - 30
Skeletonema costratum	10°-27°	2,500-5,000	15 - 30
Nannochloropsis oculata	20°-30°	2,500-8,000	0 - 36
Parvlova viridis	15°-30°	4,000-8,000	10 - 40
Tetraselmis subcordiformis	20°-28°	5,000-10,000	20 - 40
Tetraselmis tetrathele	5°-33°	5,000-10,000	6 - 53
Chlorella ellipsoidea	10°-28°	2,500-5,000	2 6- 30

Table 4.3. Temperature, light and salinity ranges for culturing various microalgae.

fore, the configuration of the culture vessel, light intensity and duration, coupled with ideal circulation patterns become more important as cell densities increase. Utilizing algae that grow mixotrophically may have a distinct advantage in intense cultures.

Temperature requirements

Microalgae that most aquarists and aquaculturists are interested in culturing are tropical strains which respond best to temperatures between 16°-27°C (60°-80°F), with an optimum of about 24°C (75°F). Small daily fluctuations within this range present no serious problems. Lower temperatures will not kill the algae, but will drastically slow growth, whereas sustained temperatures above 35°C (94°F) kill most microalgae. Table 4.3 lists suitable temperature, light intensities and salinity ranges for several species of microalgae used in aquaculture (modified from Chen 1991, Okauchi 1991). Ukeles (1961) compared temperature effects on the growth of several algae species to a control at 20°C (Table 4.4).

Algal Species	No Growth	Growth less than control	Growth equal to control at 25.5°C	Growth less than control	No growth
Monochysis lutheri	8°-9°	12°	14°-25°	27°	29°-35°
Isochysis galbana	8°-9°	12°	14°-22°	24°-25°	27°-35°
Phaeodactylum tricornutum	--	--	8°-24°	--	29°-35°
Dunaliella echlora	8°-9°	--	12°-35°	--	39°
Platymonas sp.		8°-9°	12°-32°	--	39°
Chlorella sp. (isolate #580)	8°-9°	12°	14°-35°	--	--
Chlorella sp. (isolate UHMC)	8°-9°	12°	14°-29°	--	32°-35°

Table 4.4. Growth response at various temperatures (°C).

Salinity

Salinity is important for culturing of microalgae as illustrated in Table 4.3, but salinity perhaps is more important to zooplankton. The marine rotifer *Brachionus plicatilis* is normally found inshore in more productive bay systems and estuaries and grows best at salinities of 10-20 ppt (1.0069-1.0145 SG). In order to obtain the highest rotifer production, it is prudent to grow the microalgae in comparatively low salinity ranges. Fortunately the most commonly used species, *Nannochloropsis* and *Isocrysis,* grow well at a wide range of salinities. *Tetraselmis,* another valuable species, grows better at a higher salinity range (20-30 ppt or 1.0145-1.0222 SG).

Chemical Requirements

Essential nutrients

Good growth of a microalgae culture is dependent on a proper balance between specific major and minor nutrients. Imbalances in these nutrients usually lead rapidly to cessation of culture growth. Exhaustion of nutrients is one of the most important factors limiting growth and controlling nutritional quality in mass cultures. Many off the shelf fertilizers contain inadequate amounts of nitrogen or unstable metals, especially iron, which can lead to retarded microalgae growth. Major nutrients are nitrogen and phosphorus for phytoplankton. Nitrate serves as an important nitrogen source for either freshwater or marine algae. Other combined forms of this element do occur (ammonia, nitrite, and organic compounds) an can be used when nitrate is not available. Utilization of nitrate by phytoplankton involves conversion into ammonia by the enzyme nitrate reductase before assimilation into cell material (Boney 1983). The natural assumption would be to use ammonia only as the nitrogen source. When using ammonia as a source of nitrogen a culture will initially reproduce quicker but after three to four days nitrate fed cultures often surpass the ammonia source. Although there is a lag period, nitrate is more stable and easier to use.

Phosphorus is another primary nutrient and in natural waters occurs in solution in both inorganic and organic forms. Phytoplankton cells accumulate excess phosphate reserves when levels are high This phase is referred to as the "luxury consumption phase" and utilize these reserves when natural sources are low (Boney 1983). These reserves enable cell growth to continue for a period after this nutrient is depleted. Organic forms of phosphorus are broken down by the enzyme alkaline phosphatase and occurs when inorganic phosphate is not available.

Silica in a soluble form is essential major nutrient for diatom cell wall construction and tubular skeletons. The silica content as a percentage dry weight is significant with some freshwater species ranging from 26 to 63% (Lund 1965). Utilization of silica by diatoms is dependent on sulphur metabolism and a shortage of this element could become limiting in culture work.

Iron in solution is normally always low except under acid or reducing conditions in freshwater habitats. In saltwater, which has a higher pH (8.0-8.2), iron in solution will be minimal. In alkaline solutions, iron and other metals often precipitate over time. Experimental results and field data clearly show that iron availability significantly influences both quantities and species composition. In "brown water" lakes, swamps and streams iron is mainly present in the ferrous state. This is due to the fact that high dissolved organic levels act as a chelating agent. In culture work chelating agents such as EDTA are used to maintain metals in solution, making them available to microalgae.

Manganese content in coastal waters is higher than open sea and like iron inflows are major contributors of levels. Like iron, manganese in solution is low but particulate forms are more common. In culture work, enrichment of manganese over normal natural levels increases growth of microalgae.

In addition to macronutrients, micronutrients consisting of trace metals and vitamins are usually required for optimal growth. Most microalgae are auxotrophic, so they are unable to synthesize all the necessary vitamins and must assimilate them from their environment. In nature, bacteria are a source of these vitamins. Thiamin (vitamin B_1), cobalamin (B_{12}) and biotin are considered essential vitamins for many microalgae (Spotte 1979). In fact, it has been estimated that about 70% of all planktonic algae require vitamin B_{12}.

Inorganic elements essential to many species of algae include: nitrogen, phosphorus, potassium, magnesium, calcium, sulfur, iron, copper, manganese, zinc, molybdenum, sodium, cobalt, vanadium, silicon, chlorine, boron, and iodine. Of these, nitrogen, phosphorus, magnesium, iron, copper, manganese, zinc, and molybdenum are required by all algae (O'Kelly, 1974).

Composition of growth media

Standard media commonly used in commercial aquaculture are listed in Table 4.5. Many modifications of these formulas are currently being used. Both media can be used in either fresh or saltwater. However, the f/2 medium is more commonly used in saltwater, whereas the Provasoli ES medium is often used in freshwater. Nutrient concentration varies based on how long the cultures are retained before harvest. For example, f/2 refers to 1/2 full strength whereas f/4 would be 1/4 full strength.

F/2 Medium (Guillard & Ryther 1962)		Provasoli ES Medium (Provasoli 1968)	
$NaNO_3$	150 mg/L	$NaNO_3$	105 mg/L
$NaHPO_4$	8.69 mg/L	Na_2 glycerophosphate	15 mg/L
Ferric EDTA	10 mg/L	Na_2 EDTA	24.9 mg/L
$MnCl_2$	0.22 mg/L	$Fe(NH_4)_2 (SO_4)_2\ 6H_2O$	10.5 mg/L
$CoCl_2$	0.11 mg/L	H_3BO_3	3 mg/L
$CuSO_4\ 5H_2O$	0.0196 mg/L	$FeCl_3\ 6H_2O$	0.15 mg/L
$ZnSO_4\ 7H_2O$	0.044 mg/L	$MnCl_2\ 4H_2O$	0.6 mg/L
$Na_2SiO_3\ 9H_2O$	60 mg/L	$ZnCl_2$	0.075 mg/L
$Na_2MoO_4\ 2H_2O$	0.012 mg/L	$CoCl_2\ 6H_2O$	0.015 mg/L
B_{12}	1.0 µg/L	B_{12}	3 µg/L
Biotin	1.0 µg/L	Biotin	1.5 µg/L
Thiamine HCL	0.2 mg/L		

Table 4.5. Composition of standard F2 Medium and E.S. Medium.

Oxygen, carbon dioxide, and pH

The rate of oxygen and carbon dioxide supply can become a limiting factor in growing large, dense cultures. Microalgae utilize carbon dioxide during photosynthesis and release oxygen.

Inproving circulation or the judicious addition of carbon dioxide (CO_2) or sodium bicarbonate (NaHCO$_3$) can prolong exponential growth. However, bubbling air through cultures containing more than 5% carbon dioxide can be inhibitory. Carbon dioxide and sodium bicarbonate both affect culture pH, so this should closely monitored to maintain optimal conditions. Extremely high density, controlled, commercial operations add carbon dioxide during the light period to enhance growth. Yet, CO_2 injection is expensive, and probably not necessary in most aquaculture applications. For aquaculture use, moderate aeration suffices, since it is not essential to obtain the absolute highest possible algae production per unit of time.

Excessively high or low pH depresses microalgae growth by disrupting many cellular processes. If nitrogen is supplied to cultures as an ammonium salt, most microalgae will selectively take up the ammonium ion, thus lowering the pH. Conversely, if nitrate ions are removed by algae, pH tends to increase, but is usually buffered by carbon dioxide supplied to the cultures. However if carbon dioxide becomes limiting, culture pH can climb as high as 11, eventually killing the microalgae. If the culture is maintained under ambient or a controlled light/dark cycle, pH elevates during the light period and drops during the dark period, tending to provide pH balance. When CO_2 is used, a common practice is to inject intermittently using a timer and solenoid valve to maintain pH between 7.5 and 8.5. It is therefore clear that nutrients, carbon dioxide, and pH all must be delicately balanced to obtain consistently high yields.

Alternative carbon sources like glucose, glycerol, lactic and acetic acid have been used with mixed results. Acetic acid not only lowers pH, it dissociates to provide a carbon source. Use of sugars often leads to excessive bacteria growth if microalgae densities are low. Since many microalgae have the ability to grow mixotrophically, utilization of mixed inorganic and organic nutrients are often used.

Bacteria and microalgae associations

The associations between microalgae and bacteria are becoming better understood. It has been suggested that the best algae growth occurs in bacteria-free (axenic) cultures, conditions that are difficult and expensive to maintain. However, as more experience with microalgae culture accumulates, it appears that many microalgae species grow best in association with bacteria. This is perhaps the result of nutrient recycling or nutrient complementation between algae and bacteria. An example is found in *Asterionella japonica* , which grows satisfactorily in culture as long as bacteria are present, but ceases to grow axenically, even when several organic supplements are provided (Kain & Fogg 1958). The diatom *Skeletonema costatum* grows poorly in nutrient-enriched sterile seawater, but does markedly better when bacteria are present (Johnston 1963). In this case, the enhanced growth effect is largely due to the production of essential vitamin B$_{12}$ by the bacteria, similar to what has been reported in rotifer culture (Hino 1993). In commercial aquaculture it is cost-prohibitive to maintain axenic cultures beyond subculture levels. Bacteria are so abundant and so closely associated with algae that removing them is probably more detrimental than beneficial.

Ionic Composition of Local Waters

The ionic composition of local water, especially hardness (calcium and magnesium levels) can affect microalgae growth. Newly inoculated cultures may take a few days to adapt to the characteristics of local water and grow slowly at first. Other indications of a water problem are algae clumping, settling, or failing to achieve exponential growth. Often these problems can be overcome by mixing deionized or distilled water with local water to start your cultures, then gradually introducing them to full strength local water. Vigorous stirring several times a day helps prevent clumping and settling. Clumping also can occur if phosphate concentrations are too high. Many problems can be prevented by avoiding cheap, unbalanced fertilizers.

Contamination

Contamination of microalgae cultures can be chemical or biological. Contamination can originate from multiple sources or just a single site. Usually the problem can be solved if you make careful observations, considering everything as a source of possible contamination. Figure 4.1 illustrates the most common sources of contamination.

Indications of chemical contamination

Indications of pending crashes due to contamination follow several patterns. If cultures fail to take off, clump together and drop to the bottom, but remain the natural color of the microalgae species being cultured, then the problem is most likely due to water composition, poor circulation, or a sudden temperature change. In this case, the cells are alive and often recover if agitation is increased. When a culture turns white within 24 hours it is probably due to residual chlorine or other dissolved chemicals in the water. It is essential that all chlorine be removed prior to inoculating algae cultures. If the culture remains lightly colored after 3 or 4 days with little growth, this is usually due to poor lighting or unbalanced nutrients. Older cultures (10 to 14 days) which become nutrient deficient often turn a muddy, light-green or yellow color and slowly clear. Efforts to recover a nutrient-starved culture at this point are usually futile.

Indications of biological contamination

Common types of biological contamination are excessive levels of bacteria, inadvertent introduction of microalgae consuming protozoa or rotifers, or competition from another species of microalgae. When a culture blooms, then after 3-7 days turns an off-color (yellow-green for green algae) and eventually clears, the most likely cause is contamination due to protozoans or rotifers. Competing microalgae, usually a blue-green species, are indicated by change to a more vibrant color. There is evidence that certain microalgae species may produce substances that suppress further growth. This phenomenon is called autoinhibition and has been reported in certain strains of *Chlorella vulgaris* (Pratt & Fong 1940). Diatom contamination is often indicated by a white or brownish crust attached to the culture vessel walls which entraps microalgae. Macroalgae contamination is indicated by the development of strands of green or brown algae attached to the culture vessel wall

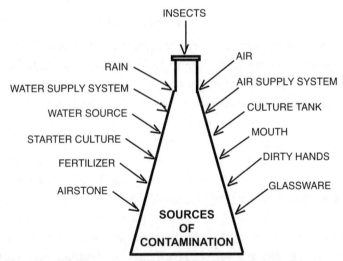

Figure 4.1. Possible sources of contamination to be considered

or free floating within the culture or changes in color. Excessive bacterial growth usually results in cloudy water, resulting in slow growth and possible collapse of the culture.

Locating contamination sources

Identifying bacterial and microalgae contamination usually requires at least 100-400X magnification. Protozoan contamination can be observed under 15-40X magnification. The most effective approach is to list all possible sources of contamination and eliminate them one by one through experimentation. Figure 4.1 illustrates the most common sources of contaminants

For protozoa, the most common route of contamination is through inadequately treated water or through the air delivery system. City water and wells can indeed be sources of contamination, so all culture water should be sterilized before use (see Treatment of Culture Water). It is also recommended that well systems be chlorinated to a 20 ppm level at least once per year. This is especially important if an underground delivery line is broken or if the system was repaired. However, even precautions will not guarantee water purity directly from the water tap.

Central Air System Precautions

A very common source of contamination is condensation in air lines which can harbor ciliates and cyanobacteria. Pockets of condensation are always present, especially when using central air systems which flow from warm, or cool moist locations into other areas with lower or higher temperatures. Locating pockets of condensation can be accomplished by using a stethoscope. It is highly recommended that air supplied to microalgae cultures originate from a separate air system which is filtered and kept dry. All air supplied should be passed through a standard clear cartridge filter equipped with a 1 or 5 μm spun, pleated, or resin filter cartridge and a constantly open bleed air valve on the bottom (Figure 4.2). All loops and sagging pipes should be supported or equipped with a drain air valves which are always open. Connections to the cultures can be equipped with air check valves to keep culture water from flowing back into the air manifolds. All air manifolds should be located above the culture to prevent water back-flow in the event of power failure.

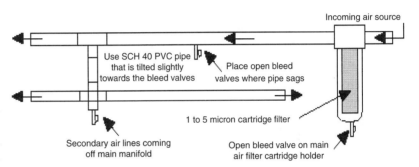

Figure 4.2 Condensation bleed valves on incoming air manifolds.

Culture Vessel Precautions

Precautions to avoid contamination in culture vessels are shown in Figure 4.3. As mentioned, a separate air source is recommended which is elevated above the level of the culture vessels. An air/water check valve may be installed, however, it not be necessary if the air source is elevated. An in-line 0.2 or 0.45 μm syringe disk filter in conjunction with a quick tubing disconnect located after the filter is a good precaution. A quick disconnect allows the culture to be easily removed from the air supply system. In-line drying tubes filled with sterile cotton and carbon are also effective. In addition, sterile cotton can be inserted into the quick disconnect cavity.

Caps should be snug, but must allow the culture to breathe. Parafilm™, mail tube caps, carboy caps, and other types can be used and modified to accommodate rigid air tubing. Polyfoam caps are not recommended unless they are used once and discarded. These tend to trap cells deep within the pores and can become a source of contamination. If an airstone is used, it should be either disposable or resistant to repeated chlorine and acid cleaning. Unique plastic diffusers that come apart in layers which can be thoroughly cleaned are available.

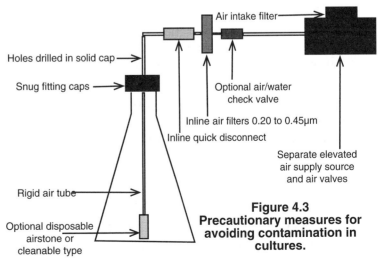

Figure 4.3 Precautionary measures for avoiding contamination in cultures.

Culture Water Precautions

Algae are sensitive indicators of poor water quality. Occasionally, high levels of dissolved metals, inorganic, or organics contaminate culture water, preventing or retarding algae growth. These usually include iron, ammonia, nitrites, nitrates, and hydrogen sulfide. Detecting specific dissolved contaminants can be a long and costly procedure. If you suspect dissolved contaminants, try starting cultures with bottled distilled or deionized water and packaged sea-salts like Forty Fathoms™ or Instant Ocean™ or Instant Ocean Reef Salts. If you have the ability to treat your own water, deionization will usually remove most metals and activated carbon filters remove most organic compounds. RO (reverse osmosis) water can also be effective for algae culture. For freshwater sub-cultures, use distilled and/or deionized water as a base. Larger freshwater cultures can be successful in some dechlorinated tap waters. Keep in mind, however, that tap water is seasonally variable and is sometimes unsuitable for algae growth. It is important that a balanced fertilizer be added to sterile water to replace the lost salts, metals, and nutrients.

Treatment of Culture Water

Pretreatment of culture water is the single most important step in culturing microalgae. When using natural seawater, lake, river, or well-water, several steps should be taken before inoculating with algae. Keep in mind that water coming out of a tap or well is not sterile. All water used in microalgae culture should be treated to remove all particulate matter and plankton including protozoans, ciliates, other algae species, bacteria, and excessive dissolved metals and organics. Treatment may be simple or complex depending on the water source used. A review of the biological treatment of culture water to improve water quality is provided by van Rijn (1996).

Common research practice is to use high pressure heat, however, this treatment often

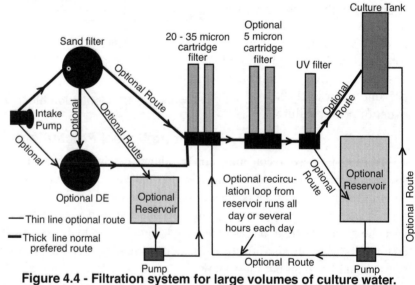

Figure 4.4 - Filtration system for large volumes of culture water.

causes precipitation of nutrients and dissolved minerals and destruction of essential vitamins. More recently, microwave heat sterilization of small volumes is being used with good success. For volumes up to about 10 gallons, heat sterilization may be preferred. However, for larger volumes heat sterilization is impractical and too expensive. For large volumes, mechanical filtration in conjunction with UV, ozone, and/or chlorination is suitable.

Nutrient media usually does not have to be sterilized if it has a low pH. Media that is acidified to pH 3.0-4.0 normally does not contain live contaminants that interfere with microalgae growth. If sterilization of nutrients is required, mechanical filtration is recommended over heat.

Mechanical sterilization

The condition of the incoming water dictates the degree of mechanical (particle) filtration needed. Large volumes of relatively clear water, can be sterilized with a simple cartridge filter arrangement in conjunction with UV and chlorine (Figure 4.4). Water with medium to heavy particulates should be filtered through a combination of a sand filter followed by an optional diatomaceous earth (DE) filter and/or cartridge filter and then ultraviolet sterilizer (UV). If properly maintained and coated, a DE filter is capable of removing particles as small as 5-10 μm. Although this is usually adequate, we prefer the final quality assurances found in using cartridge filters

Figure 4.4 shows optional pathways to filter incoming water with moderate to heavy particulates. Gross filtration is accomplished by sand and/or DE filters. From this point water can be pumped into reservoirs or staging tanks (first reservoir in Figure 4.4) capable of handling enough water for one day's use. The following day this water is passed through more refined cartridge filters and a UV sterilizer and can be either passed directly into the culture tanks or retained in another reservoir. Water in the second reservoir can be chlorinated and recirculated back through the cartridge filtration system until the next day. The following day this water is dechlorinated and added to the culture tanks. In the case of large culture tanks, the filtered water can be passed directly into the culture vessel then chlorinated and dechlorinated.

We do not fully recommend using a separate holding reservoir to retain sterilized water for eventual transfer into culture vessels unless a closed-loop, chlorinated, delivery pipe is installed. This would consist of a pump attached to the reservoir which forces water from the reservoir through the delivery lines and back during chlorination. Problems consistently occur in the transfer of sterilized water through

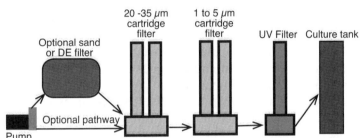

Figure 4.5 - Filtration system suitable for under 20,000 gal.

unsterile pipes, hoses, and pumps to culture vessels. We now recommend particle prefiltration, 30 μm + 5 μm followed by UV sterilization and direct transfer into the culture vessels followed by chlorination and dechlorination or ozone (see Figure 4.5).

If you should desire to store treated water, the storage vessel should be a clean, dark glass or polycarbonate capped container located in a cool, dark location. Treated water is usually not stable for very long and should be used as soon as possible after treatment. Water storage containers as well as pumps, hoses and water lines should be frequently cleaned and sterilized. If sterilized water is transferred from a reservoir to a culture vessel, it is extremely important that sterile delivery lines are used. Dark, high quality vinyl tubing is suggested as opposed to rigid PVC pipe. As a precaution, delivery lines should have shut off valves on the end, remain filled, and be flushed for several minutes before filling the culture vessel. Make sure no residual chlorine remains in any of the plumbing.

Smaller volumes of water, < 20,000 gal (75 m³) or less, can be easily handled using cartridge filters followed by a UV sterilizer. Rainbow Plastics ™ manufactures simple inexpensive modular type cartridge filters and UV sterilizers that can be easily scaled to the volume and flow required. Their largest standard unit is capable of filtering 4,500 gal (17 m³) per hour at 30-35 μm. These can be easily plumbed in parallel to handle unlimited amounts of water. Cartridge filtration offers a very wider range of options compared to standard sand or DE filtration. Special retention chambers are available that fit a standard Rainbow™ cartridge holder, into which carbon, zeolite, or specific resins can be placed for removal of excess iron, organics, and minerals. Pleated and spun polyester cartridges are available in 30, 20, 5, and 1 μm, and ultra-membrane cartridges are available down to 0.25 μm. Depending on incoming water clarity, finer filters down to 5 μm will not lower flow rates significantly. If incoming culture water is high in particulates, a sand filter should be used prior to the cartridge filter (Figure 4.5).

For even smaller volumes, 200 gal or less, a wall mounted ultrafiltration unit can be used. These units use a combination of particulate and carbon block filters followed by an ultra membrane filter capable of removing particles as small as 0.05 μm. As an added precaution, a UV sterilizer is recommended after particle filtration (Figure 4.6). Flow rates of the smaller wall mounted units are about 100 gallons (375 L) per day at 20 psi. Since carbon is used in these systems, it is recommended that nutrients be added after sterilization.

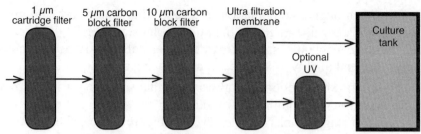

Figure 4.6 - Ultrafiltration for small volume cultures.

Vacuum filtration is another option for even smaller volumes. It is best suited for subcultures. In this case, the water is passed through a 0.45-0.25 μm filter membrane using a vacuum pump and special vacuum flask and funnel (Figure 4.7). Although this method is excellent for subcultures, it is time-consuming and requires a special filter unit, vacuum pump, and vacuum flask. Vacuum filtration is advantageous in that no precipitation or heat degradation occurs as in heat sterilization. If there is a concern of contamination or destruction by heat of nutrient media, then vacuum filtration is recommended. Fertilizer medium can be "cold-filtered" to retain full vitamin levels. Even for large volumes of culture water, it is advisable to filter sterilize the fertilizer medium to protect the vitamins and prevent nutrient precipitation.

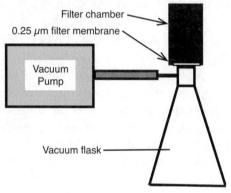

Figure 4.7 - Vacuum filtraton configuration

Water high in dissolved organics and iron is usually yellow or dark brown in color. This color can be removed by filtration through activated carbon or exchange resins. Color filtration should be done after particulate filtration and before UV sterilization or chlorination.

A cheap alternative method of particle filtration for small to medium volumes is use of 1 μm polyester felt filter bags followed by UV sterilization and/or chlorination. In this case the water is fed into the bags and allowed to gravity flow out into a collecting vessel. Then the water is pumped through a UV unit into the culture vessel and further treated with chlorine. These polyester felt bags can be repeatedly cleaned and chlorinated and come in two sizes, 18" and 31" (45 and 77 cm) long with a 7" (18 cm) diameter stainless steel ring support.

UV and ozone sterilization

Either UV or ozone can be used to sterilize water. However, optimal results are obtained in conjunction with good mechanical filtration. For maximum kill, water that is passed through UV sterilizers must be clear and free of particulates. Therefore, mechanical filtration at 30-35 μm or lower should be used prior to UV sterilization. In contrast to UV, ozone can be used effectively in colored water, but it also should be free of particulates. UV does not impart any hazardous by-products, regardless of its kill power. However, ozone at high levels can produce chloramines that are toxic at low concentration. A rule of thumb with ozone is that if you can smell it, then the output is too high and potentially hazardous to your health.

When using UV, simply decreasing the flow rate provides longer contact time within the chamber, resulting in higher kill power. The killing wavelength is 254 nm, but the effective dosage varies with what you want to kill. The flow rate is the single most important factor in achieving proper UV sterilization (Escobal, 1992). If the flow rate is 500 GPH and the diameter is 2 inches, a 40-watt mercury vapor bulb would deliver approximately 11,530 mw-sec/cm^2. If increased to a 3" diameter, the output increases to about 17,530 mw-sec/cm^2. In order to obtain dosages at different flow rates, use the following calculation (New dosage = dosage at 500 GPH x 500/new flow rate).

By cutting the flow rate by half, the dosage for a 3" diameter UV unit increases to about 34,340 mw-sec/cm^2. This dosage is more than sufficient to kill bacteria, yeast, some mold spores, viruses, and microalgae. To kill protozoa (paramecium = 200,000 mw-sec/cm^2) the flow rate would have to be decreased to about 40 GPH for a single 40-watt bulb. Based on ideal wavelength penetration, a UV sterilizer with a 3" (7.5 cm) diameter chamber is considered optimal. This diameter is up to 50% more efficient than smaller or larger diameters. See Hoff (1996) for further discussion of UV and ozone.

Heat sterilization

Culture water heavy in particulate matter should be prefiltered mechanically before heat sterilization. Like mechanical sterilization, many options are available for heat sterilization. For large volumes, over >300 gallons (1125 L), heat sterilization is probably not suitable unless solar energy is used and water is retained in an insulated tank inside a greenhouse. Using this method, a large volume of water may get to only 38°C (100°F). But if this temperature was maintained for several days, the water would be essentially pasteurized. Pasteurization by an 8-10 hour exposure to 50°C is also effective. For volumes of less than <300 gallons (1125 L), glass-lined hot water heaters or immersion heaters could be used to pasteurize the water for 24-48 hours at lower temperatures. Another form of pasteurization consists of raising the culture media to 73°C (164°F) for 10 to 15 minutes, letting cool overnight to ambient temperature, then repeating the process in 24 hours (Treece and Wohlschlag 1987). Often heating within the culture carboy can be accomplished using a 500-to 1000- watt immersion heater. Heated water should always be allowed to cool before adding nutrients or algae.

Heat sterilization under pressure (autoclaving) is the most common practice for volumes to 5 gallons (19 L). A simple inexpensive pressure cooker is easily capable of maintaining 20 psi at 120°C. A rule is to maintain this level for 15 minutes/liter and for each additional liter add 1.5 minutes. Because of the high temperatures, CO_2 is liberated and the pH rises during autoclaving, which usually causes precipitation of inorganic nutrients from the growth medium. The precipitates consist of metals, carbonates, and phosphates whose loss sometimes can be detrimental for culturing certain microalgae species. It is very important to always add the precipitate to the culture water when inoculating. Adding sterilized growth nutrients to culture water after it has cooled following heat sterilization helps eliminate precipitation.

Microwave sterilization

For over three years we have been experimenting with the use of microwaves to sterilize small volumes of culture water. This method is recommended for subcultures because it is easy, inexpensive, quick, and efficient. Ideally the culture water should be pre-filtered for particles (5 or1 μm), and then microwaved for 8 to 10 minutes per 1 to 1.5 liters using a 700 watt unit. Nutrients may or may not be added prior to microwaving since the temperature does not get over 84°C (181°F) and does not cause excessive precipitation. As suggested, the fertilizer media can also be filter-sterilized separately or pH-adjusted, and added after the culture water is sterilized. Experimentation by Bellows and Guillard (1988) showed that using a 1.2 ft^3 700-W microwave on high power would effectively kill microalgae in 5 minutes, bacteria in 8 minutes, and fungi in 10 minutes in 1.5 liters of filtered and unfiltered saltwater. If your unit does not have a rotating table, items being sterilized may have to be rotated half way through the process depending on the shape of the vessel. Microwaves equipped with a minimum of 700 watts, a high chamber profile, and turntable are preferred. Empty culture vessels, flasks, etc., also can be sterilized, but we suggest that a small amount of water be present and poured out after treatment. The complete kill of bacteria spores in dry material in a microwave or convection dry heat oven takes 45 minutes.

Chlorination

Chlorination is the most common and simplest method for sterilization, especially for large cultures. City water usually arrives at the tap with 1-3 ppm chlorine, but it may be necessary to raise the chlorine level to 20-25 ppm to kill some stages of protozoans and enteric viruses. Unlike other forms of sterilization, contact time is relatively short, varying from 5 to 20 minutes. To help reduce high chlorination levels, prefiltration of 1 to 5 μm is recommended. At low pH there are more free chlorine radicals which are highly reactive. However, at higher pH there is more combined chlorine residual such as chloramines which require a longer reaction time. At pH 6-8 it takes at least 25 times more combined available chlorine to produce the same germicidal efficiency (Banerjee & Chermisinoff, 1985). Chlorine test strips or common swimming pool test kits are useful for determining the presence of free chlorine ions. Although the strips are designed for freshwater, they still provide a yes or no color reaction in saltwater, but cannot be used to determine concentration.

Chlorine disinfection of culture water can be accomplished with liquid household or pool bleach which is made with sodium hypochlorite at concentrations of 5% to 15% available chlorine. A 2.5 ppm concentration is recommended as a starting point and can be obtained by using 1/2 ml of fresh bleach per liter of water. However, liquid bleach solutions lose potency with time and should be checked periodically. In addition, liquid chlorine mixtures, including household bleach, contain stabilizers (e.g. curanic acid) which help retard chlorine loss. Consequently, even after 24 hours of aeration, highly toxic chlorine levels may still be present and more than the usual amount of sodium thiosulfate (a chlorine neutralizer) may be necessary. We do not advocate use of liquid bleach because of these problems.

Dry granular pool chlorine or pool shock-treatment are most reliable for small and large scale treatment of culture water. Dry chlorine mixtures containing calcium or sodium hypochlorite are effective. Either of these forms dissolves readily in water and contains approximately 70% available chlorine (Banerjee & Chermisinoff, 1985). Levels of 12-25 ppm are recommended for raw septic sewage and 1- 5 ppm for sand-filtered effluent. Dosage of 1 oz (28 gm) per 500 gal. (1875 L) yields a 1-3 ppm concentration. A more practical measurement is 1/8 teaspoon (0.60 ml) per 5 gallons (19 L) of culture water. However, unlike liquid bleach, the potency decreases quickly when mixed with water since there are no stabilizers. Proper dry storage is absolutely necessary to maintain potency.

To chlorinate culture water, first add the appropriate amount of dry chlorine to warm water, stir, and allow to dissolve as much as possible. Pour the dissolved liquid portion only into the culture water. Under no circumstances should you add any of the insoluble particulates to the culture water. Passing the chlorine solution through a small mesh brine shrimp net or 50-100 μm netting can remove undissolved particulates. After adding the dissolved chlorine, provide heavy aeration for 12 hours. In virtually clear water, actual kill time usually is 10 to 60 minutes at 20-25°C depending on pH. For example, at a pH of 6.0 available free chlorine residual after 10 minutes is 0.2 ppm, and combined chlorine residual after 60 minutes is 1.0 ppm. However at a pH of 9.0, free chlorine residual at 10 minutes is 0.8 ppm, and at 60 minutes the combined chlorine residual is >3.0 ppm. At higher pH more neutralizer may be required.

Be careful of contaminants in water you sterilize with chlorine. Waters with ammonia can form chlormines when chlorinated. Free chlorine is a very active oxidizing agent and is therefore highly reactive with readily oxidized compounds such as ammonia. The formation of chloramines greatly reduces chlorine's effectiveness in killing bacteria and viruses (Kohler, 1953) and higher levels of chlorine may be needed as ammonia levels increase. Three forms of chloramines exist, mono-, di-, and trichloramine. The specific reaction product depends on the pH of the water, temperature, time, and initial chlorine-to-ammonia ratio. In general, monochloramine and dichloramine are generated in the pH range of 4.5-8.5. Above pH 8.5 monochloramine usually exists alone, and below pH 4.4 trichloramine is produced (Banerjee & Chermisinoff, 1985). Chloramines are common in urban water treatment and cannot be removed by aeration, thus sodium thiosulfate or activated carbon must be used to dechlorinate. Sodium thiosulfate solutions are available under several commercial names from pet stores or chemical supply houses, or it can be obtained in dry form from aquaculture suppliers.

Most waters are sufficiently chlorinated after 30 minutes exposure to 2.5 ppm chlorine, provided no ammonia is present. However, we prefer to extend exposure to 12 to 24 hours to achieve a margin of safety. As a precaution, add a small amount of sodium thiosulfate and mix for an additional 15 minutes to assure dechlorination. Check for residual chlorine using a test strip before inoculation. Approximately 175 mg of sodium thiosulfate per liter is sufficient to dechlorinate a 2.5 ppm solution. A one molar solution of sodium thiosulfate is 153 gm/liter distilled water. Be careful with sodium thiosulfate since an excess may reduce trace metal availability. It is good practice check the chlorine level with a simple swimming pool chlorine test strip before inoculating with algae just to make sure dechlorination was successful. Even if you cannot smell chlorine, enough still may be present to retard algae growth.

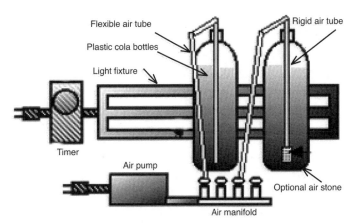

Figure 4.8 - A simple microalgae culture system utilizing easy to obtain cola bottles as culture vessels.

Equipment

For small scale or hobby level cultures most of the equipment needed can be easily obtained from local sources. We have attempted to keep the equipment and process as simple as possible, but it still requires that you carefully follow good culture procedures. Figure 4.8 illustrates a simple phytoplankton setup using items found at home, your local hardware store or aquarium shop. Figure 4.9 is a self contained culture unit manufactured by Florida Aqua Farms Inc. which utilizes disposable plastic bags as culture vessels. Figure 4.10 illustrates a commercial sized system for growing microalgae and zooplankton using fiberglass microalgae culture cylinders and plastic sinks for the zooplankton.

Container shape and size

Many different shapes and sizes of vessels can be used to culture microalgae, from rectangular tanks to plastic buckets or suspended plastic bags. However, the shape of the culture vessel strongly influences circulation patterns and light exposure, so it is especially important to optimize container shape in order to achieve high cell densities. A cylinder shaped container with concave or flat bottom, clear transparent sides, and a restricted or covered opening is ideal.

For small scale, home aquarium-level cultures we feel that glass, wide-mouth gallon jars or clear plastic cola bottles are suitable (Figure 4.8). If glass jars are used, it will be necessary to make a plastic top equipped with a small hole for the air line. Glass jars can be reused after thorough cleaning. Probably the most convenient container is a plastic cola bottle. The bottom of the bottle is round providing optimal circulation and the neck is narrow which reduces contamination. Best of all, they are easy to obtain, disposable, recyclable, and cheap.

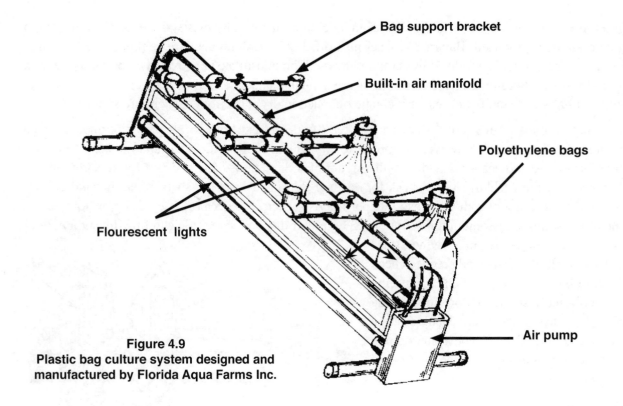

Bag support bracket

Built-in air manifold

Polyethylene bags

Flourescent lights

Air pump

Figure 4.9
Plastic bag culture system designed and
manufactured by Florida Aqua Farms Inc.

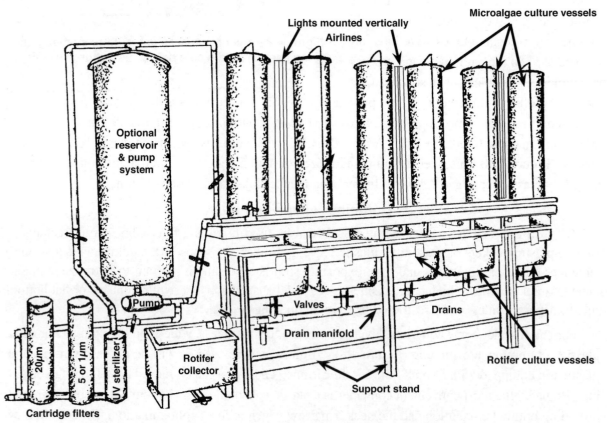

Microalgae culture vessels

Lights mounted vertically
Airlines

Optional reservoir & pump system

20µm

5 or 1µm

UV sterilizer

Pump

Rotifer collector

Valves

Drain manifold

Drains

Support stand

Rotifer culture vessels

Cartridge filters

Figure 4.10 - Small commercial-size culture unit for microalgae and zooplankton.

For small scale cultures beyond cola bottles and up to 5 gallons (19 L), clear plastic or glass water jugs are very suitable and reasonably priced. However, their restricted mouth makes them hard to clean. Florida Aqua Farms offers an inverted plastic jug (Figure 4.11) which has had the bottom removed to allow easy access. A plastic valve also is permanently mounted for easy draining. These have been successfully utilized in our facilities for years.

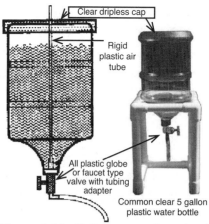

Figure 4.11 - Culture vessel made of a converted plastic carboy bottle manufactured by Florida Aqua Farms

If opaque buckets, circular or rectangular tanks are used, they should have bright white interiors to gain optimal light reflection. In addition, good vertical lighting must be provided. If glass aquaria are used three sides should be covered with aluminum foil and lights should be mounted on the open side. Circulation is very important in square culture vessels. Use of more than one air outlet and daily stirring may be necessary to eliminate pockets of sedimentation. Clear plexiglas, glass, or stretch plastic covers should also be used to reduce contamination.

Cleaning culture equipment

Equipment including air lines, airstones, culture vessels, measuring devices, air manifolds, etc. should be routinely cleaned with a good detergent and light acid bath. Most biodegradable dishwasher or lab detergents are suitable, but rinse them thoroughly with hot water. After rinsing, soak and brush all surfaces for a few minutes with a 5% hydrochloric acid solution to remove attached diatoms and other contaminants. A dilution of muriatic acid, which is available from pool supply stores, can also be used. More recently, we have been using hardware store grade denatured alcohol for sterilization by soaking or washing down counter tops, tubing, filters, etc.

Heat sensitive culture vessels should be rinsed with very hot water, drained, dried upside down, then sealed with aluminum foil or plastic wrap until used. If you are using heat resistant glass or high density plastic culture vessels, we strongly recommend damp subculture vessels be microwaved for 3 minutes or placed in a pressure cooker at 15 psi for 20 minutes. After cooling, cover mouth with aluminum foil, plastic wrap, or Parafilm™ for storage.

Aeration

Aeration devices should always be installed above the water levels of all culture vessels to prevent back siphoning. A separate air source is highly recommended for algae culture. Preferably, it should have a filtered intake and output to help avoid contamination (Figures 4.2 & 4.3). Air supply lines should be made of clear, clean, flexible vinyl tubing connected to rigid plastic or glass tubes. Only rigid tubes should be used inside the culture vessel. Cut rigid tubing two inches or longer than the depth of culture vessels and bend at the top by using heat. This bend helps prevent the air line from collapsing, and stopping the air supply. Quick disconnects placed close to the output valve allow easy removal of the culture vessel without removing the airline and disrupting the vessel seal. Small pieces of sterile cotton can be placed inside the disconnect for further filtration. Use of airstones is optional. If used, fused glass-bead or plastic types that can be taken apart are recommended since they must be cleaned repeatedly in mild acid solutions. Medium to coarse bubbles are recommended since fine bubbles create foam and strip algal cells from the water. Air manifolds should be made of plastic and be capable of being taken apart and cleaned. Interconnecting plastic aquarium air valves are ideal for constructing manifolds.

Aeration of subcultures is not aways necessary. However, faster growth and higher densities are achieved with aeration. Non-aerated cultures are typically placed on a shaker table for continuous mixing or manually swirled one or two times a day. Use of aeration enables cultures to grow faster, but more contamination precautions must be employed.

Lighting

Most microalgae use red and blue light for photosynthesis, so lighting should consist of fluorescent bulbs emitting 400-700 nm wavelengths. "Cool White™" and common fluorescent bulbs are usually strong in the red range, but weak in the blue. Vita-light™ is a bright white, more natural light that has a strong broad spectrum and provide more appropriate wavelengths for algal growth. "Gro-Lux™" bulbs are strong in the red, while actinic type bulbs like are strong in the blue. The best results are obtained using a combination of white light and actinic bulbs. Bulbs should be changed every 12 to 15 months depending on condition of the bulb, and growth response of the algae. More recently, metal halide lights have been used, but are more effective on large cultures of 500+ gallons (1875 L), grown in opaque-walled tanks. Halide lights from 75 to 1000 watts are normally used depending on the size and depth of the culture. However, they generate considerable heat, and should be located at least 12" (30 cm) above the surface. It has been suggested that additional blue fluorescent bulbs in conjunction with halides provide optimal growth.

We suggest artificial lighting over sunlight since the duration and intensity of the latter cannot be easily controlled and may cause overheating or insufficient light, reducing culture growth. However, if natural light is used, cultures should be placed near windows receiving primarily morning sun. If cultures are maintained in a greenhouse, you may want to install 40% shade cloth on the west side of the building during the summer to reduce heat. If the building is positioned lengthwise in a north-south orientation, a strip of 60% shade cloth down the middle of the building will help during the summer months.

Fluorescent light fixtures should ideally be mounted about six inches away, either horizontally or vertically, depending on the height of the culture vessels (Figures 4.8, 4.9 & 4.10). However, if mounted above a water surface, protection for the bulb and fittings should be provided. Lighting can be controlled by a simple light timer and set for a minimum of 16 to 24 hours light. Depending on your particular conditions you may get better results using continuous light. However, photo-inhibition is possible if the light is too intense for too long. Small containers, high temperatures, and nutrient deficiencies increase the chances of causing photo-inhibition. Even though artificial lighting is more easily regulated it is also a costly option in large operations.

Aquarium salts

If you are culturing marine microalgae, it is important to select a good quality aquarium salt mixture that is well balanced, like Instant Ocean™, Forty Fathoms™ or special new blends such as Reef Crystals™. Rock salt commonly used in water softeners, or table salt are not suitable because their ionic compositions do not resemble seawater. Since most microalgae are found in bay systems and coastal waters the salinity should be in the range of 12-27 ppt (1.015-1.020 density). Do not purchase a large container of salt unless you intend to use it within 1 to 6 months. A small box that can be resealed is best, or you can store a portion of the newly opened salts in an airtight plastic container. Most brands contain sodium thiosulfate for removing chlorine which is found in most city tap water. All salt mixtures go through chemical changes when mixed, and must be stabilized and balanced before starting a culture. To be on the safe side, mix salts and heavily aerate for at least three to four hours before adding the fertilizer and microalgae. Best results with all artificial salts are obtained by mixing a concentrated solution in a blender for 2-3 minutes and then diluting to desired salinity. For subculture work we prefer mixing salts in distilled water. However, standard sterilized bottle water is suitable. We do not advocate natural seawater for subcultures because of seasonal variability however for large volumes there is no other economical choice.

Fertilizer

We recommend a fertilizer specifically formulated for microalgae and based on Guillard's f/2 enrichment medium. This formulation has yielded good results for many species of algae for over twenty years. Premixed f/2 media are available from several sources including Micro Algae Grow™ & Microalgae Grow Mass Packs™ (Florida Aqua Farms Inc.) and f/2 media (Fritz Chemicals). Other fertilizers may grow microalgae, but the quality and value of microalgae is directly influenced by the properties of its aquatic environment. We strongly advise you not to use common terrestrial plant fertilizers, especially those containing ammonia or urea as a nitrogen source. Residual ammonia or urea from overfertilized cultures or those that are harvested prematurely may contain toxic levels of ammonia. In addition, common agricultural fertilizers usually do not contain essential trace metals and vitamins suitable for good microalgae growth and are often high in phosphates.

In mass cultures, 10,000 gallons or larger, fish emulsions and/or trace metal solutions in conjunction with sodium nitrate and phosphate can be used. When fertilizing large volumes of water, liquid mixtures are more reactive and controllable than using slowly decaying organic media. Standard practice in ponds is to introduce organic media such as cotton seed, oat and rice meals, chicken manure, and other dehydrated organic products in dry form and allow them to slowly decay (Geiger 1983, Li et al., 1996, Ludwig, 1997). Experiments have shown that mixing these ingredients in a separate tank with aeration and allowing them to digest into a slurry and then introducing them into a pond produces more positive results and quicker blooms (see Daphnia Chapter).

Growth Patterns of Microalgae Cultures

Five growth phases are typically recognized (Figure 4.12):

1) Lag or introduction phase where little increase in cell density occurs. This phase is easily observed when an algae culture is transferred from a plate to a liquid culture. Keep in mind how physiologically difficult it must be for normally planktonic, motile microalgae like *Isochrysis, Tetraselmis* and *Dunaliella* to make the transition from liquid culture to an agar plate and vice versa. Plate cultures of microalgae are forced to survive under cold, dark, nutrient-limited conditions in a semi-dormant state. Many metabolic enzymes become inactivated and concentrations of cellular materials fall to levels preventing cell division. Therefore, it takes a short period of acclimation for algae to adjust to their new, aqueous environment before resuming growth.

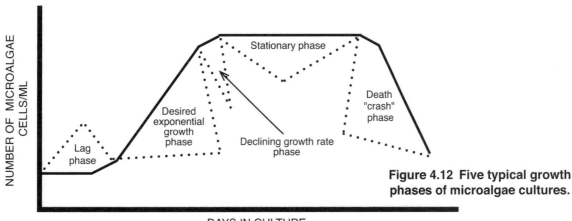

Figure 4.12 Five typical growth phases of microalgae cultures.

Specific dissolved minerals and nutrients may increase or shorten lag phase. Depending on the algae species, high levels of calcium, magnesium, or phosphorus may extend lag phase. Another factor contributing to lag phase is the requirement for threshold concentrations of specific excreted dissolved compounds to be reached before exponential growth begins. Metabolic synthesis of these compounds to threshold concentrations is called preconditioning of the medium. Glycolic acid is a compound that is excreted into the medium by many microalgae species (Tolbert & Zill 1956; Whittingham & Pritchard 1963).

2) Once lag phase is completed, cultures enter an exponential growth phase where cell division increases rapidly according to a logarithmic function. Cells harvested during the upper limits of this phase are nutritionally balanced and preferred. Inoculating new cultures from exponentially growing populations is recommended and can substantially shorten the time required for up-scaling. Duration of the exponential growth phase can be controlled by daily harvesting (1/3-1/2 of the culture volume) and nutrient replenishment, or by transferring the culture to larger vessels with new culture media (Figure 4.17).

3) Duration of exponential growth varies depending on nutrients, light, carbon dioxide, culture vessel, or other physical and chemical factors which limit growth. Towards the end of exponential growth phase cell division slows significantly. Cultures in this phase can be recovered but, the window for recovery is small. Cells harvested at this phase are at their peak in nutritional value and density. When using batch cultures, cells are harvested during this phase (Figure 4.15 & 4.16).

4) Stationary phase occurs when cell densities remain relatively constant for an extended period of time, with alternating periods of decrease followed by periods of increase. In batch cultures, nutrients are consumed without replacement and stationary phase may be very short if cell densities are exceptionally high. In continuous cultures this can be extended by low levels of nutrient replenishment. Cells harvested from this phase are not as nutritious as those taken during exponential growth.

5) As water quality deteriorates and cells starve, cell density begins to decrease rapidly, and a situation called a "crash" develops. Crashes can occur for several reasons including, exhaustion of a nutrient, O_2 deficiency, or a pH disturbance. Culture crashes can happen overnight, so it is important to closely monitor cultures for early warnings of a crash like declining cell densities or changes in culture color. Once the culture changes color, recovery is virtually impossible. Algae cells from declining populations also have little nutritional value for zooplankton.

The best advice when starting an algae culture is to have patience. Many factors affect algae growth including: physiological conditions of inoculant cells, water quality, light, temperature, fertilizer, and pH. Algae cells must acclimate to the prevailing conditions before rapid cell division commences.

Obtaining Microalgae Cultures

Microalgae inoculants

Two types of microalgae starter cultures are available: those that are axenic (Greek = without foreigners) containing no bacteria and unialgal cultures which consist of one species of algae with associated bacteria. For aquarists and aquaculturists, unialgal starter cultures are normal and perfectly fine for inoculating mass cultures since these are not grown in totally sterile conditions. Unialgal starter cultures are less expensive than axenic cultures and are available from various universities, laboratories, and Florida Aqua Farms. Inoculants come in several forms including liquid cultures, agar slants, and unique agar petri dish cultures (Micro Algae Disks™, Florida Aqua Farms). These are lightweight, inexpensive to ship, and can be stored in a cool, dark place for months. The alga most commonly provided on disks and liquid cultures is *Nannochloropsis*. Several common saltwater microalgae species also are available on disks including *Isochrysis*, and *Tetraselmis*. Microalgae preferred for freshwater applications are *Nanochloropsis, Chlorella* and *Nannochloris* of which all can be grown in either fresh or saltwater.

Starting your own inoculant

If you wish to use a particular alga found in your area then there are several ways algae cells can be isolated. These methods are time-consuming and take patience, since you are working with microscopic plant cells and attempting to grow a culture from very few isolated cells. Isolating specific algae can be achieved by several methods including dilution, pipetting, and streaking on agar plates (Belcher, 1982).

Dilution

Dilution (Figure 4.13) consists of removing a sample from a pond or seawater which is dominant in two or three species of algae. A portion is divided into five test tubes with 5 ml of sterile culture media. One of these is further divided into an additional 5 test tubes with 5 ml culture media etc. for a dilution of 4 or 5 times. These are allowed to grow for 6 weeks or longer. Usually a few of the middle dilutions will consist of unialgal cultures. If this is not the case, those with fewer cells are further divided.

Isolating one algae cell for cloning by pipetting is difficult, since one has to work at 100X magnification and have patience. A sample of water is placed on a sterile slide without a cover slip. Another sterile slide is placed in a covered petri dish and single drops of sterile culture water are arranged in a row. A disposable glass pipet with a dispensing rubber bulb on the end is heated along the tip to soften the glass. A pair of tweezers are held on the end of the tip. When hot enough, pull the heated glass with the tweezers to form a very fine diameter capillary tube on the end of the pipet. Snap off the end with the tweezers. Under the microscope, locate the desired algae cell. Place the thin capillary tube into the drop containing the algae sample. By capillary motion the cells are sucked into the tube. Quickly remove the tube and force the sample into one of the sterile drops on the other slide. This is repeated until several cells are in one drop. From this point the slide with multiple drops is examined under the scope and the drop containing the isolated cells is further subdivided into other drops until only the desirable cells are retained in one or more drops. If single cell clones are wanted then the drops must be carefully examined. Drops containing the desired cells are then transferred into sterile culture test tubes with 5 ml sterile media. The transfer pipet must be heat or alcohol sterilized between transfers.

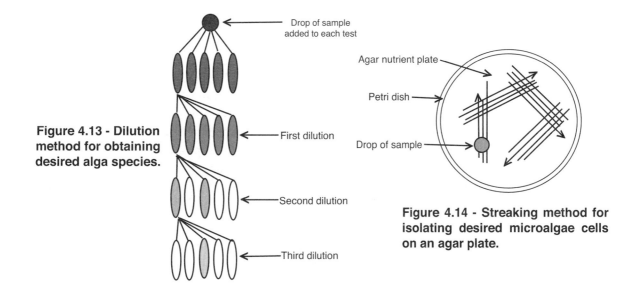

Figure 4.13 - Dilution method for obtaining desired alga species.

Drop of sample added to each test

First dilution

Second dilution

Third dilution

Agar nutrient plate

Petri dish

Drop of sample

Figure 4.14 - Streaking method for isolating desired microalgae cells on an agar plate.

Agar Streaking

Streaking (Figure 4.14) consists of spreading the sample over nutrient agar in a petri dish. Nutrient agar plates can be made by adding Guillard's f/2 or Micro Algae Grow™ to hot agar and pouring a 1/4" deep layer into a covered petri disk. Once the agar has cooled, place the sample on the plate and draw 4 or 5 parallel lines through the sample with a sterile wire. Then draw several lines originating from the end of the first lines, repeat until the plate has several sets of parallel lines on the surface. Place cover back on the dish and allow sample to grow for several days. Small circular colonies of algae will develop along where the lines were drawn. Isolated colonies are picked from the agar surface with a sterile tool and placed in liquid culture media. Timing how long the plates should be allowed to grow before removing colonies is essential to reduce merging and overall contamination by bacteria, molds, and protozoans.

Agar Plating

Similar to plating and streaking procedure above where clumps of cells and also isolate cells grow on the agar medium. A 2% solution of agar mixed with a suitable mineral medium (marine or fresh water) is autoclaved and poured into petri-dishes. The algae sample is then removed using a wire loop and streaked across the surface in a zigzag pattern. Place the dishes upside down below a light source. Algal cells with associated bacteria will show a cloudy halo around them and should not be used. A clean patch of cells are selected, picked up with a sterile loop and streaked on a second agar plate so further refining. The cells are then transferred to a liquid culture. Agar slopes (1.0 - 1.5% agar solution) in screw top test tubes can also be used instead of petri-dishes for long term storage of cells. Hot sterile agar is poured into sterile tubes and allow to dry on an 30 to 45° angle which provides and larger surface for streaking the cells on the agar. The tubes are then capped and kept at room temperature and in low to medium light.

Filtration

Filtration screens (plankton cloth) are available as small as 0.5μm to 25μm which covers the size of a wide range of algal species. Due to the fine nature of these screens they are normally not adequate for filtering large volumes of algae unless the filter is designed with a rotating washed drum screen. For example you can small drum-shaped diatoms like *Thalassiosira pseudonano* (size 11-14 μm) from a small circular algae like *Nannochloris atomus* (1.5-2.5 μm) using a 5μm screen. Using this technique screens trap all the larger cells and about half of the smaller specie (Sournia 1978) Colony formation and cell protuberances, as well as cell size and shape, influence retention.

Phototactic Behavior

We have adapted several other methods based on how algae respond to light and salinity. *Tetraselmis* and *Isochrysis* are motile and phototactic (attracted to light) which are features that allow separation from nonmotile algae like *Nannochloropsis* and *Chlorella*. *Tetraselmis* tend to be more phototrophic then *Isochrysis*. Pour the contaminated algae culture into a clean clear glass flask. Provide no air to the culture vessel and allow to settle for a day or two so the nonmotile algae settle to the bottom of the flask. Take a small flashlight and concentrate the light toward the top of the vessel since *Tetraselmis* and/or *Isochrysis* are motile they concentrate around the light. Using a fine disposable pipet equipped with a dispensing rubber bulb very carefully remove cells in the water column and not on the glass and place in low volume (25 ml) of sterile culture water with no aeration. Slowly step up the culture to 100-150 ml over a couple of weeks before providing aeration.

Salinity Tolerance

Salinity tolerance can also be used to separate several species. For example, *Tetraselmis* tolerates significantly higher salinity them many other algae. Simply raise the salinity 5-10 ppt per day to about

50 to 60 ppt and use the light technique above but further down in the water column since some non-motile algae may float in higher salinities. If you want to remove *Tetraselmis* and/or *Isochrysis* from a nonmotile algal like *Nannochloropsis* and *Chlorella* which can tolerate freshwater, lower the salinity 5-10 ppt per day to 0 ppt salinity and culture the nonmotile algae for several weeks then start to raise the salinity. To separate *Tetraselmis* from *Isochrysis*, two motile algae, you might want to try to raise the salinity since *Tetraselmis species* tolerate higher salinities.

Eliminating bacteria for axenic subcultures

Boney (1983) outlined a procedure for obtaining axenic cultures (no bacteria in culture). As a general rule washing of plating methods are preferred over use of antibiotics for reducing bacteria in cultures. Use of antibiotics is the easiest to use but has the disadvantage that undetectable cell injuries can take place. It is harder to culture antibiotically cleaned cells and often other nutrient medias may have to be used to force them to grow. Many nonmotile phytoplankton organisms and filamentous algae with muci-lage cover can only be cleaned with antibiotics (Boney 1983). The procedure uses high dosages of a broad spectrum antibiotic over a short time. Most algae are more tolerant of high antibiotic concentrations then are bacteria.

A freshly prepared antibiotic mixture (12-15 ml) is filter-sterilized and 6 ml is used per series of six tubes containing dense, vigorous algal growth. Following the procedure in Figure 4.15 culture A" with 6 ml of culture is selected then diluted in half with 6 ml of antibiotic mixture.

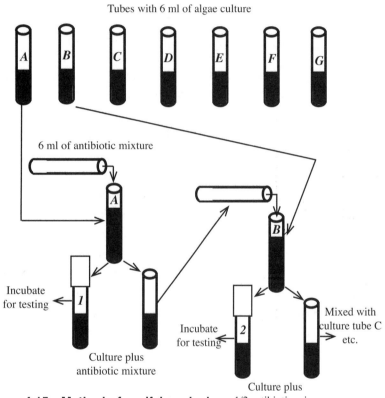

Figure 4.15 - Method of purifying algal cultures with antibiotics (Droop 1967).

This mixture is then divided and one is marked as "1" and set aside for testing. The other is then used to dilute culture "B" and then divided equally into a treated culture marked "2" for testing and the other for diluting culture "C" etc. After all six algal cultures are inoculated with diluted levels of antibiotics, six treated tubes are created. After this procedure one drop of *sterility test medium* is then added to encourage growth of bacteria in the subcultures over a 24 hour period. Minimal bacterial contamination will become evident as a cloudy appearance of the water. Sterility test medium is a sterilized solution made with distilled water and bacto-peptone preparation (0.1%). After the 24 hour treatment new subcultures are made using antibiotic free media. In general it is suspected that tubes 5, 6 and possibly 4 still contain bacteria and tubes 3 to 6 and possibly 2 would contain living algae. Generally tubes 3 and 4 are the best choice for further axenic subcultures.

The recommended antibiotic mixtures are listed below.

Antibiotic	For Diatoms	For Flagellates
Benzyl penicillin SO4	8000 μg cm-3	8000 μg cm-3
Strretomycin SO4	1600 μg cm-3	2000 μg cm-3
Chloramphenico	l200 μg cm-3	8 μg cm-3

Maintaining Liquid Stock Subcultures

Once you have obtained or started a liquid culture of desired microalgae, you must continually retain a clean source for inoculating new cultures. In laboratories this is done in culture tubes (20 x 150 mm) with screw tops, sterile cotton plugs or loose covers. However for continuous aquaculture purposes larger starter cultures consisting of 250 to 500 ml flasks are more suitable.

The number of subcultures set up depend on the number of mass cultures you are going to start on a daily or weekly basis. We prefer 2 subcultures for each larger culture you are going to set up. For example, if you are going to set up 2 five-gallon (19 L) mid-cultures per week, then you should have a minimum of 3 to 4 subculture flasks that are from 7 to 14 days old for inoculation. One will be used to inoculate each 5 gallon carboy and the others serve as backups and/or a restart for new subcultures. Below is a modified classical method used in laboratories.

1. Into culture tubes or flasks place desired volume of clean, f level (regular strength) fertilized, culture water. Replace cap loosely or place cotton plug (open tubes only) in opening.

2. Sterilize by autoclaving (15 min. at 20 psi) or microwave (7 minutes). Note, black screw caps can only be microwaved for 2 minutes. However if submerged in water they can withstand the 7 minute duration. Allow to cool to room temperature, then tighten caps or cover with Parafilm™ until needed. Fertilizer that is acidified to pH of 3-4 does not require heat sterilization and may be added after sterilizing the water.

3. Disinfect the surface of the work area and hands with denatured alcohol or other suitable disinfectant. Turn off air-conditioners, blowers, fans, or other circulating devices.

4. Agitate old stock by swirling (do no shake) and allow larger particles to settle. Once settled, remove cap and flame the open end of culture tube or rub with denatured alcohol.

5. With a sterile pipet or pouring, remove enough of the old culture to create a density that you cannot see through, flame culture or alcohol clean tube mouth again and replace cap loosely or cover with Parafilm™. However, if the purpose is just to keep a culture alive for future use then inoculants can be smaller.

6. Remove cap from newly sterilized culture tubes and flame or clean mouth. Drip inoculum from pipet or pour into the new medium, flame or clean mouth, and replace cap loosely or cover with Parafilm™.

7. If the purpose is to just maintain a culture for future use, then a static, non-agitated, culture tube or 150-250 ml flask is suitable. Agitate daily by swirling and place in a test tube rack or shelf by a medium to low 24- hour light source. Keep subcultures sufficiently away from larger cultures to lessen the chance of contamination.

8. Repeat this procedure weekly, biweekly, or monthly depending on demand and growth. Retain older stocks in reserve until the next month's cultures are started. Start larger cultures with subcultures that are one to two weeks old.

9. The volume and density of test tube subcultures are only suitable to culture a 150 to 250 ml subculture, which in turn is suitable to inoculate a 1 to 2 liter culture vessel. In lieu of liquid

starter tube cultures, plated algae cultures, called Micro Algae Disks™ (Florida Aqua Farms) or fresh algal paste can be substituted.

10. If the purpose is to maintain large cultures on a continuous basis, then agitated subcultures of 500 ml to 1 liter are more suitable and greatly reduce labor (Figure 4.16). Agitated subcultures would be installed as shown in figure 4.18. Inoculation levels should be dense to the point where nothing is distinguishable when looking through the flask.

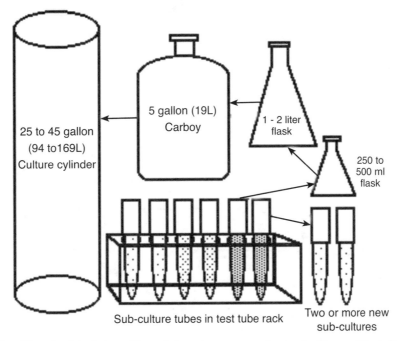

Figure 4.16 - Classical batch culture laboratory procedure starting with culture tube sub-cultures. Usual growth period for the test tube stage is 14 to 30 days, 250-500 ml flasks is 7-10 days, 1-2 L flasks is 5 to 7 days, carboy 19 L stage is 5 to 7 days, and the final growth period 5 to 7 days.

Procedure for Algae Culture

Classical Batch Culture Method

Three methods are typically used to culture algae: batch, modified batch, and semi-continuous. The classical batch method of microalgae culture consists of inoculating culture tubes with a low density of algae cells. After a week or two, test tube cultures are transferred into 250 ml flasks and later into larger carboy, 19 L (5G) culture vessels. (Figure 4.16). This process of up-scaling is time consuming because of the extended lag phase associated with starting from small inoculants. It is important to match algae production with zooplankton food requirements because overproduction wastes time and money, while underproduction leads to inadequate larval feed.

Modified Batch Culture

Our suggestion for the typical aquaculture facility is the modified batch culture method outlined in Figure 4.17. It consists of eliminating the standard test tube inoculants and substitutes larger, high density 500 ml cultures as inoculants. Although liquid tube cultures, slant agar cultures, or Micro Algae Disks™ may be maintained as backups, they are not used as the primary inoculant. In addition, this modified batch method is usually more predictable and easier to control than continuous cultures.

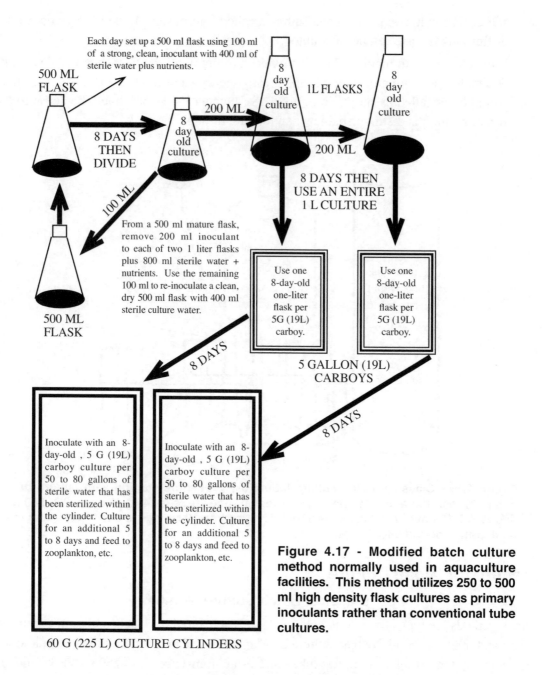

500 ML
FLASK

Each day set up a 500 ml flask using 100 ml of a strong, clean, inoculant with 400 ml of sterile water plus nutrients.

8 DAYS
THEN
DIVIDE

8 day old culture

200 ML

8 day old culture

1L FLASKS

8 day old culture

100 ML

200 ML

8 DAYS THEN
USE AN ENTIRE
1 L CULTURE

From a 500 ml mature flask, remove 200 ml inoculant to each of two 1 liter flasks plus 800 ml sterile water + nutrients. Use the remaining 100 ml to re-inoculate a clean, dry 500 ml flask with 400 ml sterile culture water.

500 ML
FLASK

Use one 8-day-old one-liter flask per 5G (19L) carboy.

Use one 8-day-old one-liter flask per 5G (19L) carboy.

5 GALLON (19L)
CARBOYS

8 DAYS

8 DAYS

Inoculate with an 8-day-old , 5 G (19L) carboy culture per 50 to 80 gallons of sterile water that has been sterilized within the cylinder. Culture for an additional 5 to 8 days and feed to zooplankton, etc.

Inoculate with an 8-day-old , 5 G (19L) carboy culture per 50 to 80 gallons of sterile water that has been sterilized within the cylinder. Culture for an additional 5 to 8 days and feed to zooplankton, etc.

Figure 4.17 - Modified batch culture method normally used in aquaculture facilities. This method utilizes 250 to 500 ml high density flask cultures as primary inoculants rather than conventional tube cultures.

60 G (225 L) CULTURE CYLINDERS

Semi-Continuous Cultures

The following procedure is designed for small scale cultures by aquarists, or for up-scaling starter cultures to large mass cultures by aquaculturists. These methods may seem unconventional to those who have spent great effort maintaining axenic cultures. Yet, these methods are practical and have been successfully applied in aquaculture for many years. The following procedure is also suitable for home aquarists. It consists of re-blooming cultures several times before cleaning the vessel and restarting cultures with fresh inoculant. This method employs less control and usually produces lower algae densities than cultures that are cleaned before each reuse (Table 4.6). However, lower cell production per unit volume can be overcome by producing more algae volume than needed on a daily basis.

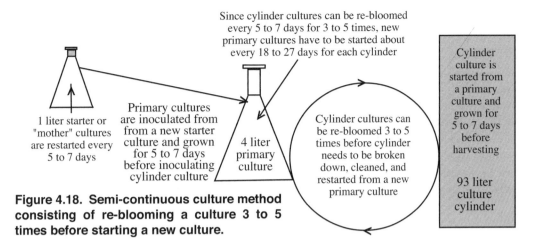

Since cylinder cultures can be re-bloomed every 5 to 7 days for 3 to 5 times, new primary cultures have to be started about every 18 to 27 days for each cylinder

Cylinder culture is started from a primary culture and grown for 5 to 7 days before harvesting

93 liter culture cylinder

1 liter starter or "mother" cultures are restarted every 5 to 7 days

Primary cultures are inoculated from from a new starter culture and grown for 5 to 7 days before inoculating cylinder culture

4 liter primary culture

Cylinder cultures can be re-bloomed 3 to 5 times before cylinder needs to be broken down, cleaned, and restarted from a new primary culture

Figure 4.18. Semi-continuous culture method consisting of re-blooming a culture 3 to 5 times before starting a new culture.

General Culture Methods

Regardless of which method is used, the same basic procedures and precautions must be followed.

1)   Cleanliness is foremost. Keep all equipment squeaky clean and away from spray and other sources of contamination. Most microalgae and many contaminants cannot be seen at less than 100X magnification. Consequently, even a wet finger or glassware can be a source of contamination for rotifers, rotifer cysts, live *Artemia, Artemia* cysts, ciliates, other protozoans, or algae species. Disposable culture containers like clean plastic cola bottles or new plastic bags are useful for aquarists and aquaculturists. Yet, in large scale cultures, like those used in aquaculture, glass Erlenmeyer, Fernback, or boiling flasks are recommended for starter cultures. Recently we have been using round bottom, glass, boiling flasks since they provide optimal circulation. All vessels should be brush washed in hot water using quality, biodegradable detergents like Calgon™ or Alconox™, rinsed in hot water, acid cleaned and brushed with a 30% dilution of muriatic acid or 5%-10% solution of hydrochloric acid, rinsed again in hot water, microwaved for 3 minutes, cooled and capped until used. Rough, white colored, interior surfaces on culture vessels indicate the presence of diatoms and/or calcium deposits which trap algae cells. These can easily be removed with acid solutions. Goggles, rubber gloves, and good ventilation are essential in addition to the safety instructions listed on the acid container. This same acid solution can be used for rigid air tubes, flexible airlines and air valves. Airstones can also be acid cleaned, but soak them no longer than 3-5 minutes. In commercial aquaculture, it is probably best to replace flexible airline routinely.

2)   Make sure all equipment is clean and dry before starting a culture. Saltwater cultures require about 20 to 25 cc aquarium sea salt mixture per liter (5 to 6 teaspoons), which yields a salinity of about 20 to 25 ppt (1.0145 to 1.0183 density). Note that a standard, heaping coffee measurer will suffice for this amount. Small volumes of seawater should be thoroughly mixed and pH stabilized by mixing in a blender for 2-3 minutes.

        Large volumes of artificial or natural seawater should be prepared in a clean, dark reservoir or preferably within the culture vessel itself. Heavy prefiltered aeration should be provided for at least 6 hr to balance the pH. Since aquarium sea salt mixture can also be a source of contaminants, it is advisable that all sources of culture water be chlorinated then dechlorinated prior to inoculation. The source of chlorine should be from dry chlorine powders at 1/8 teaspoon per 5 gallon (19 L). Refer to "Treatment of Culture Water" for more details on pretreatment of natural or recycled aquarium water.

If reservoirs of pretreated water are used, they should be made of plastic or glass and kept in a cool, dark area. Black plastic sheeting or black paint around the outside of the reservoir will prevent algae growth. A reservoir should be only big enough to hold 1 to 3 days of culture water. Therefore, more than one reservoir is advisable to prevent delays in starting up a culture. Contamination via transfer equipment from the reservoir to the culture vessel is always a risk and should be a priority consideration.

3) After the culture water has been mixed and pretreated, the fertilizer can be added. Care should be taken when selecting organic-based fertilizers, since some could be sources of pathogenic bacteria. Some find it necessary to sterilize the culture water and fertilizer together, but for most applications careful selection and handling of fertilizer solutions usually will yield contaminate-free cultures. Heat sterilization of fertilizers can greatly reduce their potency, causing phosphates and some metals to precipitate and inactivate vitamins. Preferred sterilization for fertilizer solutions is microfiltration and acidification (see Treatment of Culture Water).

Dosage varies according to how dense the inoculant is and how long you are going to culture the algae. If you are using Micro Algae Grow™, a modified Guillard's f/2 medium (Florida Aqua Farms), add 1/3 to 1 ml per liter (1/16 to 1/4 teaspoon per quart). Be cautioned that adding more fertilizer does not produce faster growth and may actually inhibit or slow growth. Best growth is achieved when macro and micronutrients are in proper balance. Initially we recommend using the lowest dosage of 1/3 ml per liter. Using too high a dosage could be detrimental to the zooplankton and/or larvae being fed the microalgae. For example, if you use the standard f/2 dosage of 1 ml/L, a fairly dense culture will grow and survive for 10 to 14 days. However, if you plan to harvest algae after only 5 days growth, then the dosage should be half or less. If the fertilizer dose is not reduced, you could accumulate excess macro and trace nutrients in the zooplankton and/or larval cultures which could be detrimental.

4) There are several ways to inoculate your culture with microalgae: liquid cultures, test tube agar slant cultures, and agar disk cultures (Micro Algae Disks™). Agar disk or slant cultures have several advantages in that they are easy to transport and can be stored for 6+ months prior to use. Disk cultures provide high cell inoculants which helps reduce up-scaling time. Liquid test tube cultures provide the quickest starting inoculant, but have a brief shelf life of 7 to 10 days.

A general rule for inoculant levels varies from 2-10% of the volume being cultured. This sounds good but does not address the density of the inoculant. When inoculating cultures of any size, it is important to provide the strongest inoculant possible to reduce competing bacteria from becoming dominant. Preferably, inoculant levels should be dense enough to totally block vision through the culture vessel. The objective of inoculation is to achieve a uniform medium-light green or brown color depending on the algae being cultured. Denser inoculations result in faster log-phase growth and are less prone to contaminant invasion by other algae and bacteria.

If using a Micro Algae Disk™, best results are achieved by utilizing the entire disk to inoculate a 500 ml to one liter (1/2-1 quart) culture. Although 9-liter (5-gallon) cultures can be started with a single disk, often the lag phase will be longer, thus providing opportunity for contaminants to become established.

When using disks or slants, the following procedure is preferred. Remove seals, flood the agar surface with 20-30 ml of sterile culture water, cover, provide moderate continuous light, and allow to soak for 12-24 hours. With a sterile cotton "Q-Tip", rub off the cells and pour into the culture vessel. Place more culture water on the disk or slant, rub the surface again and flush off remaining cells. Some algae have a tendency to clump and flake off the agar. Soaking agar plate cultures helps reduce this and enhances the initial growth phase.

If you have difficulty in establishing a freshwater culture from an agar disk using the method above, it is sometimes helpful to blend the entire disk in a blender with some culture water. With a clean knife lift the entire nutrient agar base out of the petri dish and place into a clean blender. Add about 200 ml (1 cup) of sterile culture water and homogenize for about 30 seconds. Avoid longer blending times because they can injure algae cells and slow growth. Add this mixture to your culture water.

5) Contamination precautions including filters should be included wherever possible regardless of the culture method used. Quick disconnects are very useful and allow easy removal of the culture vessel without disruption or removal of cap seals or aeration tubes (Figure 4.3). In addition, larger versions of quick disconnects allow you to insert a small piece of sterile cotton inside. In-line filters or drying tubes can be filled with carbon and/or sterile cotton, then mounted before the air manifold. The top of the culture vessel should be closed with stretch wrap, Parafilm™, cotton, or a snug plastic cap with a hole for insertion of the air tube. Note, cultures must be able to "breathe" so caps and seals should be snug, but not to the point where air cannot easily escape.

6) Moderate aeration should be provided via a rigid plastic tube equipped with or without a glass bead airstone (Figure 4.18). Better circulation can often be achieved by eliminating an airstone but, is dependent on the shape of the culture vessel, circulation pattern, and the algae being cultured.

7) Culture vessels should be placed about 15 cm (6 inches) away from the light source. If using incandescent light, it may be necessary to locate cultures further away to prevent overheating. If using sunlight, locate cultures in a window with northeastern exposure, preferably one that does not receive direct sun. Light duration should be from 14 to 24 hours.

8) A 2 to 4 liter culture will reach maximum cell density in 7 to 10 days, depending on inoculant level, temperature, and nutrient composition. Larger cultures may take longer but higher initial inoculant levels help reduce up-scaling time.

9) At this point, the entire culture can be harvested and another culture started using a portion of the harvested culture (see Figure 4.16).

10) The objective of a continuous culture is to maintain a culture in exponential (log) growth phase. Using the following method, a large culture can be maintained in log growth phase for several weeks. Begin harvesting about 3/4 of the culture on day 5-7 when cells are still in log growth and restock the culture vessel with sterile, fertilized water. This dilutes accumulated metabolites that can inhibit microalgae growth and resupplies nutrients. Staggering several cultures will allow harvests every day. Another technique normally used on large cultures >374 L (>100 gal) is to remove 1/3 to 1/2 daily and replace with new sterile culture water and the appropriate level of nutrients. Keep in mind that excess nutrients in the culture water can be detrimental to zooplankton, therefore, nutrient levels in continuous cultures should be closely monitored.

11) Unless quite sophisticated closed system methods are employed, continuous cultures utilizing the same vessel are usually suitable for regrowth only two or three times. Cell density and quality are likely to decrease with each reinoculation and the probability of contamination increases. Since small volume cultures are often used as starters to inoculate larger and more

expensive mass cultures, it is prudent to limit how long a single liquid culture is utilized as inoculant. This is best done experimentally under the culture conditions prevailing at your site.

12) The size of inoculant required for up-scaling a culture is based on several considerations. Strong inoculations typically yield the best results with minimal initial lag periods. We have found that 3 liters of *Nannochloropsis* at a density of 8-15 million cells per ml is sufficient to inoculate a 25-gallon (95-liter) culture. This inoculates the culture at approximately 220-515 thousand cells per ml, a 27-fold dilution. At this inoculation level, cultures of 3-8 million cells per ml can be harvested in 5 days at 25°C. A rule of thumb in inoculation is never to dilute more than 100-fold. Greater dilutions usually result in cultures failing to enter log growth phase. Typical growth curves of re-bloomed and new 95 liter cultures are shown in Table 4.6. Cylinders inoculated with new 3 liter starter cultures typically yielded about 1 million less cells per ml after 5 days than re-bloomed cultures.

Day	New Culture	Re-bloomed Culture
0	0.38 ± 0.15	0.68 ± 0.28
1	0.89 ± 0.65	2.15 ± 0.87
2	2.26 ± 1.56	3.36 ± 1.48
3	4.01 ± 1.79	5.25 ± 1.27
4	5.05 ± 1.07	6.20 ± 1.35
5	5.25 ± 0.75	6.03 ± 1.76

Table 4.6. Average algae densities (millions of cells/ml) in 25 gallon (95 L) started from new stock cultures versus re-bloomed cultures. Numbers shown are mean ± standard deviation.

Estimating Algal Cell Densities

Algae grow very rapidly in a culture, and when growth stops, several metabolic pathways are initiated that may decrease nutritional quality. In order to maximize the amount and quality of microalgae, it is essential that microalgae population growth be closely followed to determine optimal time for harvesting. Once a protocol has been established that works for your facility, less monitoring is needed.

Direct Counts

Direct counts are the techniques used to examine and count the number of algal cells present per unit of volume. For microalgae, a compound microscope is essential with powers from 100X to 400X. The number of cells counted are usually expressed as cells/ml.

Hemocytometer

Hemocytometers, (Figure 4.19) were originally designed to accurately count human blood cells, has been used to count small microalgae for years. A hemocytometer is a thick glass slide with a mirrored surface which has precisely etched grids defining a known volume. A special cover slip is placed on top of the mirrored surface and a drop of the algae sample is added. The sample is drawn under the coverslip by capillary attraction.

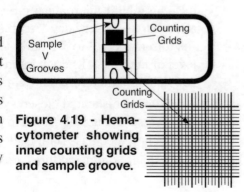

Figure 4.19 - Hemacytometer showing inner counting grids and sample groove.

To determine the density of algae cells, counts are made at 100X magnification and multiplied by a factor supplied by the manufacturer (typically 10,000). If your sample is too concentrated, it may be necessary to dilute the sample to the make counting more accurate. Although this method takes more time than the Secchi depth, it is substantially more accurate and permits checking for contamination while quantifying algal density.

Motile microalgae

If you are using a motile microalgae it is easier to kill the cells before counting by using the following procedure.

1. Place a well mixed 1 ml (equals 30 drops if using a fine-tipped transfer pipet) culture sample in a deep well porcelain plate.

2. Add 1 drop of 5% formalin and mix well, then go to #4.

Nonmotile microalgae

3. If you are using a nonmotile microalgae it is not necessary to kill the cells.

4. Place the long dimension of coverslip horizontally parallel to the long dimension of the slide, so that both central counting grids are covered but the V shaped grooves are exposed for easy access.

5. With a fine-tipped transfer pipette mix culture samples well and then place one drop in each V-groove until the culture is drawn under the coverslip. Note, some models do not have grooves. Hemocytometers without grooves are loaded by pipetting the sample onto each grid and placing the coverslip on top.

6. Most hemocytometers are divided into twenty-five 1 mm squares in 5 rows by 5 columns (25 mm^2 grid, Figure 4.18). Each of the 1 mm squares is subdivided into sixteen 0.25 mm squares. Using 100X magnification (the 10X objective), count the number of cells in the central twenty-five - 1 mm squares. A hand counter is useful for counting algae cells. Using a compound microscope on the 100X magnification (10X objective), count the number of cells in the central 25 squares which are each further divided into smaller squares.

7. For the highest degree of accuracy it is best to count all the cells in both sets of grid (there are 2 sets of grids on a hemocytometer), average the two counts, divide by 0.98 and multiply by 10,000. While counting a hand counter is recommended.

8. This figure is the number of cells in one milliliter (ml) of your microalgae culture.

9. For a quick, less time consuming count, but with less degree of accuracy, you can do the following:

a. If it is a low density culture it is best to count all 25 squares in one set of grids and then multiply by 10,000.

b. If it is a medium density culture count about 5 squares of one grid set, multiply this by 5 then multiply by 10,000. If there are too many cells to count, count the cells in the most representative 1 mm square. Multiply by 25 to obtain the number of cells in the entire 25 mm^2 grid then 10,000 for cells/ml. High densities can also be diluted by a factor of 10 or 100, then multiplied accordingly.

10. While counting, check the sample for any contaminants like ciliates, protozoans, diatoms, or filamentous algae. Use a 40X lens for checking for blue-green algae contamination. If you are using a motile algae you may want to check a sample without killing the algae since most contaminants are motile and easier to see.

11. Carefully remove the coverslip and gently clean the surface of the hemacytometer and coverslip with tissue or lens paper. Avoid scratching the slide surface with the coverslip or tissue. Heavy duty coverslips designed specifically for a hemocytometer are available at an additional cost. However, we have used standard, cheap, 24 x 40mm covers without significant count discrepancies.

12. Transfer pipettes must be flushed well between samples with distilled water. Often a light chlorine bath is recommended while not in use.

Nannoplankton Chamber

The Palmer-Maloney nannoplankton counting chamber also requires a count of the number of cells present with a microscope. Instead of a mirrored-surface that is used in the hemocytometer, the chamber is designed to hold exactly 0.1 ml (Figure 4.19). The number of algae in the entire chamber is counted and multiplied by 10 to calculate the number of cells per ml. A less tedious way is to count one or more individual microscope fields and average the number of cells counted. To determine the area of your field (mm^2) using the formula: $p \times r^2$, where r is radius of the field in mm (i.e. 1/2 the diameter of the field) and p is the pi constant 3.14159. The chamber has a defined depth of 0.4 mm when the cover slip is in place. The formula for calculating the volume is (area)x(depth).

The nannoplankton counting chamber has several advantages over the hemocytometer: 1) It is a thinner slide that will allow examination and counting of microalgae at higher magnification, 2) large cells will not be crushed by the weight of the slide, 3) it is less expensive and easier to use.

To use this chamber, place the cover slip over the space in the slide and leave a small space to add the sample. As the water fills the chamber, it will draw the cover slip over the chamber. The cover slip then is positioned to seal in the sample, avoiding trapping any air bubbles. Allow the sample to settle for approximately two minutes, then the number of algae present per field or in the entire chamber is counted. If cells do not settle, adding Lugol's solution to the sample before placing it in the chamber will increase the settling rate.

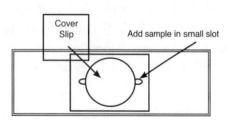

Figure 4.20 - Palmer-Maloney counting chamber.

Indirect Counts

These techniques also quantify the number of algae present using parameters other than direct examination through a microscope. They are useful and rapid, however are usually more expensive and need to be calibrated initially by using one of the direct techniques.

Coulter Counter

This is an electronic particle counter that can be used to determine the number as well as the size of cells in a sample. As the sample is drawn through a small aperture, the cells in the sample cause a small change in an electrical field. This change in electrical potential is measured and can be related to the number of cells passing though the aperture. Once this machine is calibrated for a given species, it is usually very reliable. Problems result from non-algal particulates (e.g. sand, debris), which get counted along with the microalgae, yielding higher cells densities than actually exist.

Spectrophotometry

Spectrophotometers illuminate samples in a special test tube using light of specific wavelengths. They measure and quantify any color substance in the range of 350-750 nm.

Chlorophyll a

The major pigment in most of the microalgae can be examined by measuring the amount of light absorbed at 664, 647, and 630 nm. The following technique can be used:

1. Measure 10 ml of microalgae.

2. Determine the number of cells per ml, using a hemocytometer or a nannoplankton chamber.

3. Centrifuge the sample at 5000 rpm for 5 min.

4. Decant off as much liquid as possible.

5. Grind cells in a tissue homogenizer in 10 ml of 90% acetone saturated with $MgCO_3$ (approximately 1 gram/liter).

6. Centrifuge again as in step 3.

7. Place a sample containing only acetone in the spectrophotometer and set the absorbance to 000 (this is the blank). Remove blank, insert the chlorophyll sample and read absorbance.

(A) at the appropriate wavelengths for chlorophyll (chl) *a*. Use the following equation calculate the amount of chlorophyll as micrograms (μg)/ml:

$$\mu g\ chl\ a\ /ml = 11.85XA_{664} - 1.54\,XA_{647} - 0.08XA_{630}$$

8. Calculate the μg of chlorophyll/cell by simply dividing the μg of chlorophyll from step 7 by the number of cells from step 2.

Turbidity

Algal cells suspended in water make the water cloudy or turbid. This turbidity can be used to measure cell number by setting the spectrophotometer at 750 nm and measuring absorbance. The greater the absorbance, the greater the number of cells present. Turbidity can be calibrated against one of the direct counting techniques. One problem with any turbidity technique is interference from non-algal sources like bacteria or quick settling. Samples should be shaken and promptly read. Standard curves may have to be developed for each species of algae.

Fluorometer

The amount of chlorophyll may also be determined with a fluorometer, an instrument that can be set up to specifically examine this pigment. The microalgae cells in a test tube are exposed to blue light. The subsequent emission of red light by the cells is termed fluorescence, and can be measured electronically. This technique is much more sensitive than the spectrophotometric technique described earlier, and can be used to determine the health of a microalgae culture. In general, as cells approach stationary phase, they tend to be more fluorescent and this parameter can be used to detect physiological changes in microalgae

cultures that occur before any visible changes appear. In addition, techniques for determining the amount of lipid are being developed for this instrument. This will become an important way of monitoring and enhancing the nutritional quality of microalgae, Berglund et al. (1987).

Secchi Depth

Secchi depth is a means of measuring water clarity in field studies of aquatic habitats. This technique also relies on the turbidity that algal cells impart as described above. Secchi depth is determined as the depth at which a standard white disc disappears from view. A convenient algal density stick which measures secchi depth is available from Florida Aqua Farms. This stick is immersed into the culture and the depth that the disc is no longer visible is recorded (see photo on page 60). The secchi depth is proportional to the cell density. Once calibrated with a direct counting technique, it is a useful estimate for monitoring a microalgae culture. Secchi depth calibrations using a hemocytometer are presented for several microalgae in Table 4.7.

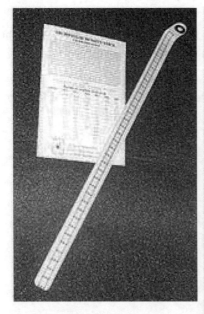

DEPTH (cm)	NCP	CHL	NAN	TET	DUN	ISO
1	27.6	44.8	96.4	2.1	12.6	30.1
2	19.2	32.4	61.4	1.8	8.7	23.2
3	13.3	23.6	39.1	1.6	6.1	17.2
4	8.8	17.1	24.9	1.3	4.2	12.9
5	5.2	12.4	15.8	1.2	2.9	9.7
6	2.9	9.0	10.1	1.0	2.0	7.2
7	2.0	6.6	6.4	0.87	--	5.4
8	1.9	4.8	4.1	0.75	--	4.1
9	1.8	3.5	2.6	0 .65	--	2.6
10	1.4	2.5	--	0.56	--	--
11	1.1	1.8	--	0.49	- -	--
12	0.69	1.3	--	0.42	--	--
13	--	1.0	--	--	--	--

Secchi Density Measuring Stick
mfg. Florida Aqua Farms.

Table 4.7. Conversion table of depths (column 1) using a modified Secchi Microalgae Density Stick. Readings were taken in clear sided 2-liter cola bottles with overhead fluorescent lighting. Cell density listed were determined by hemocytometer counts (columns 2-7). Cells/ml are obtained measuring the depth of when he white dot disappeared, selecting the algae you are using, and then multiplying the table numbers by 1,000,000. The algae are: NCP- Nannochloropsis, CHL- Chlorella, NAN- Nannochloris, TET- Tetraselmis, DUN- Dunaliella, ISO- Isochrysis.

Concentrating Algal Mass

Microalgae is normally feed directly from the culture in liquid form however, recently there is interest in using "algae paste" or concentrate forms of microalgae. The concentrate is stored in the refrigerator and diluted when used. Utilization of a paste form drastically reduces the physical space required to rear micoalgae and is attractive for many applications. However, nutritional quality is always a concern during a extended period of time. This concept is not new to aquaculture but recently several new companies are now offering concentrated preserved and fresh forms of microalgae. Concentrated algae paste from your cultures can be obtained using several methods. Various techniques are reviewed by Fox (1983) and Barnabe (1990). For small scale harvesting filter screens of cartridge filters (1μ to 5μ) can be used. Concentrated cells can be washed off with limited amounts of water then directly used, refrigerated, or preserved.

Flocculation

Chemical flocculation is often used. Utilization of this process is centered around clarification of sewage and industrial waste. Chemicals such as aluminum sulphate, ferric chloride, Sodium silicate, magnesium chloride, calcium carbonate and many more are added to dense cultures causing cells to coagulate and precipitate out of solution, accumulating on the bottom or surface. Coagulation and precipitation results from neutralization of electrically charged particles. Natural organic agents such as gelatin, chitosan, sodium alginate and clupeinecan also be used. Although chemical flocculation is easy to apply to mass situations such a ponds, consideration of which chemical is best depends on how the algae is going to be utilized. Hills and Nakamura (1978) outlined five considerations when selecting a suitable method for food purposes: (1) harmless to humans, (2) dosage is minimal, (3) resultant shall have neither high acidity or alkalinity, (4) taste or coloration is not compromised, and (5) price for procedure is minimal. Siphoning or skimming is used to harvest coagulated cells. Coagulated cells can be fed directly to detritus bottom feeders such as shrimp and other species.

Centrifugation and filtration

Large centrifuges can handle large volumes of algae but efficiency may be a problem. Hills and Nakamura (1978) reported that this equipment is capable or removing 20% of the culture whereas co-agulation and filtration can remover 95% of the algae. Recently high speed rotating particle drum filters have shown promise. Culture water is poured onto the filter drum trapping the algae on the surface while water passes through the screen. Algae paste is then removed by scrapping or a light washing.

Resulting algae-paste is then used directly or preserved for later use. Additions of cryoprotective preservatives such as glucose and dimethylsulfoxide are used to maintain cellular structure during freezing. Not all algae species respond the same to these treatments and considerable experimentation must be conducted to find suitable methods for long term preservation. For example concentrated *Tetraselmis suecica* just kept in darkness at 4° C (39° F) maintain viability for several weeks but, when frozen viability is lost. Hermetically sealed vials of concentrated cells lose their viability more rapidly than those kept in cotton-plugged vials (Coutteau, 1996).

Trouble Shooting

Occasionally, disaster strikes and cultures will crash or are difficult to start. Most failures are due to accidently introduced contaminants which consume and/or inhibit growth (see Contamination Section). A crash does not always imply the algae is dead or beyond reviving. Cells that clump and/or fall to the bottom yet retain their normal color, the crash is probably due to an imbalance of dissolved nutrients, pH, water changes, or significant rapid temperature changes. If algae cells are bleached white, yellow, or very pale, the problem is likely due to toxic chemicals or zooplankton contamination.

Although there are several ways for a culture to fail, many problems are often easily remedied. The following are some tips for corrective action.

1) Using a central air system in which air lines are exposed to hot temperatures in one area and cool in another will lead to condensation in the lines. Water buildup in air manifolds can become easily contaminated with blue-green algae, protozoans, or bacteria which do not need light to survive and can be easily passed into your cultures. Air entering the culture room should be passed through a clear filter housing equipped with a 1-5 μm cartridge with a bleed valve on the bottom. This can be followed with U-shaped trap for collecting moisture, equipped with a constantly open air valve. Routine flushing of the entire air manifold with chlorine bleach or an acid solution will also help prevent contamination. Ideally, a separate air system equipped with an intake filter is best for the culture room.

2) Contaminated nutrient sources can be a source of failure. It is important to store opened nutrients in a cool, dark, dry place. Refrigeration is preferable. If contaminated, nutrients should be replaced immediately. To test for contamination, add nutrients to sterile culture water as usual, but eliminate the algae inoculant, wait for five days and see if anything blooms.

3) Dirty culture vessels can also lead to failure. If you are using the same vessels over and over, it is important to use a good quality detergent such as Alconox, acid dip and brush, rinse

thoroughly, microwave, and allow to dry. Glass vessels should be clear, have no rough inner surfaces, and be "squeaky clean." If the glass surface is rough, acid wash. Many culturists use disposable plastic bag liners which eliminate problems associated with inadequate cleaning.

4) Dirty airstones, airlines, and airline weights are always a problem and should be acid washed or replaced after each use.

5) Cultures can be contaminated with zooplankton. *Artemia* and rotifer cysts can be a real nuisance, since they can be easily transferred in either a wet or dry state. Phytoplankton cultures located too close to aquaria or other sources of zooplankton can be contaminated by spray. To help prevent spray transfer, use a plastic divider between microalgae and zooplankton cultures and make sure that algae cultures are capped.

6) Bacteria, microalgae cells, and some protozoans can be transported through the culture room on air currents. Air conditioners which do not drain properly can harbor contaminants which can be transported in the air. Prevent air currents that are directed towards your cultures. Make sure air conditioner filters are clean and the units drain properly.

7) If you experience early crashes or your culture never starts, there are several other factors to consider.

a) Culture water may not be aged and or aerated long enough to remove water treatment chemicals or dissolve all salts.

b) There could be a problem with the mineral or metal composition of your water supply. If you suspect this, try cultures prepared with deionized or distilled water.

c) If a saltwater culture, make sure your salts are dry and clean or sterilized if wet. Artificial seawater should be vigorously mixed with no residual undissolved salts. Several hours of vigorous aeration or 3-5 minutes in a blender usually accomplishes this. Try another brand of seasalt for comparison.

d) If using continuous culture, check each inoculant for contamination with zooplankton, protozoa, or other microalgae. Many zooplankton can be observed at 8-15X magnification. With strong lighting and no magnification, large zooplankton often can be seen as white particles with darting movements. Microalgae contaminants require a 100X magnification or sometimes can be detected by color changes in the culture.

e) Nutrients could be too old and have lost potency however, this is usually not the case if they are derived from inorganic ingredients.

f) Over fertilization can be a problem especially with certain motile algae such as *Tetraselmis* and *Pyramimonas*. An overdose does not usually kill the algae but cause them to drop to the bottom and become non-motile. Stirring cultures several times a day will help alleviate this problem.

g) Too low an inoculant per volume or other factors mentioned allow bacteria to proliferate and often subdue a new culture. This will appear as a whithish cloud in the water. If this occurs just stir the algae several times a day. When the algae is dense enough the bacteria   will be subdued.

h) Your microalgae inoculant could be contaminated. Inoculants are unialgal before opening, but once opened and stored, they can become contaminated, especially if you are using only a portion of the inoculant. If you use a liquid culture, keep in mind that many cultures currently sold in aquarium hobbyist markets are mixtures of several microalgae, bluegreens, or other species which may not be suitable as a food source.

i) Culture vessel may be toxic or unclean. It is possible for certain containers to be toxic to microalgae, or the container could have been used previously for something toxic.

j)   Algae clump and will not remain suspended in the water column. Calcium or magnesium concentrations could be too high; excessive phosphorus may be present. Re-make culture water exactly according to directions using a microalgae fertilizer. Try utilizing another base water source like bottled distilled water.

k)   Insufficient or too much light or improper wave lengths.

l)   Culture overheated or too cold. Quick temperature changes can often cause microalgae to drop from suspension. This does not mean the cells are dead or that they cannot be revived. Cultures often resume growth once they acclimate to the new temperature as long   as it is not too extreme.

8)   A late crash is usually due to low nutrients and perhaps in some cases insufficient light as the culture becomes denser. Foam indicates cells are dying, excess dissolved organics are given off, and are being striped from the water by aeration. Also too fine an air bubble may cause premature foaming and cell loss.

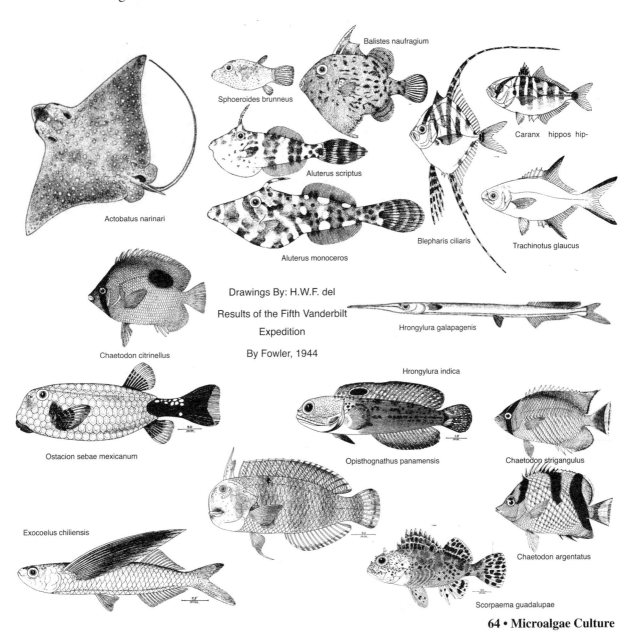

Balistes naufragium

Sphoeroides brunneus

Aluterus scriptus

Aluterus monoceros

Actobatus narinari

Caranx   hippos hip-

Blepharis ciliaris

Trachinotus glaucus

Chaetodon citrinellus

Drawings By: H.W.F. del

Results of the Fifth Vanderbilt

Expedition

By Fowler, 1944

Hrongylura galapagenis

Hrongylura indica

Ostacion sebae mexicanum

Opisthognathus panamensis

Chaetodon strigangulus

Exocoelus chiliensis

Chaetodon argentatus

Scorpaema guadalupae

Chapter 5 - ROTIFER CULTURE

Introduction

Rotifer means "wheeled animal" because of the wheel-like feeding appendages which appears to rotate. The phylum Rotifera consists of 3 classes, 120 genera and approximately 2000 described species ranging in size from 100 to 2500 μm (microns = millionths of a meter = micrometer). Most species live in freshwater or semi-aquatic habitats like in mosses, but some live in coastal marine habitats and salt lakes . Rotifers may account for >50% of the zooplankton biomass in a freshwater lakes at particular times of the year. Most species are holoplanktonic (free swimming) herbivores, but some are bacteriovores or predators. They generally move by swimming or crawling, with some species sessile as adults, permanently attached to plants. For over 100 years rotifers were regarded as pests in Japan. Eel culture was one of the early successes in Japanese aquaculture, but a phenomenon called "mizukawari", a summer phytoplankton kill, threatened to limit the industry. Mizukawari was a serious problem because the loss of phytoplankton populations disrupted the ecological balance of the eel culture ponds leading to a rapid degradation of water quality. Oxygen content of pond water typically was lowered to less than 1.0 mg/L (ppm) during mizukawari, causing death of the eels due to asphyxiation. By the early 1950's, eel growers had enough and a scientific investigation was launched into the causes of mizukawari. Beginning in 1955, Dr. Takashi Ito and colleagues published a series of nine papers detailing the mizukawari process and some possible remedies. Their work demonstrated that mizukawari was caused by dense populations of filter-feeding rotifers that consumed virtually all of the phytoplankton in the ponds. These voracious zooplankters were therefore considered noxious pests to be eliminated from eel cultures. In fact, in Ito's first papers were instructions for treating ponds to selectively kill rotifers.

As intensive aquaculture was initiated for other species besides eels, it became clear that larval rearing was a major limitation for many marine fish, crabs, and shrimp. Ito and his colleagues turned the rotifer problem into a solution by collecting rotifers from eel ponds and feeding them to larval ayu fish, Plecoglossus altivelia. Their experiment was a resounding success and rotifers proved to be a superior larval feed for several other fish and crustaceans, such as the prawn Penaeus japonicus and the crab Portunus trituberculatus. Suddenly there was great interest in culturing instead of killing rotifers. Ito (1960) wrote a paper that has become a cornerstone for rotifer mass culture. Over the past 40 years there has been a great deal of research has been completed on rotifers and their use in aquaculture (Lubzens at al. 2001). In Japan it is recognized that the fry production of many fish species is usually limited by rotifer production (Hirata, et al. 1983; Fukusho, 1989a and b; Fukusho and Hirayama, 1992). Over 70 species of marine fish and 18 species of crustaceans have been reared using live rotifers as a larval food source. Consequently, there is currently great interest in the biology of rotifers and how it can be manipulated for aquacultural production.

Besides these practical considerations, using rotifers as a food for the culture of marine larval fish and invertebrates has a sound ecological rationale. In marine environments, rotifers are most abundant in estuaries where they are the numerically dominant zooplankters during certain seasons. Most rotifers are small herbivores less than 400 μm (microns = millionths of a meter = micrometer) in length and consume phytoplankton 3-17 μm in diameter. Larval stages of many marine fish and invertebrates are spent in estuaries where rotifers, because of their size, slow swimming behavior and digestibility are easy prey for larval predators. As a result of this natural ecological association which has evolved over millions of years, many marine fish and invertebrates are well adapted to capture and nutritionally utilize rotifers (Houde and Zastrow, 1993).

In the past several years, significant advances have been achieved in rotifer culture technology, nutritional enrichment and biomass storage. Rotifer strains of different sizes also have been identified that are easier for small, first feeding larvae to handle. Breakthroughs in manipulation of the rotifer life cycle for cyst production have introduced additional possibilities for rotifer culture. The primary objective of this chapter is to familiarize aquaculturists with recent advancements in rotifer culture techniques so that they can be incorporated into routine culture procedures.

Several topics are considered, including: the role of rotifers in larval rearing, the rotifer life cycle, rotifer culture - physical and chemical requirements, food requirements, containers and systems design, rotifer production, storing rotifer biomass, improving the nutritional quality of rotifers, variability among strains, and sources of materials. Use of rotifers in aquaculture has mainly been as a larval feed, with the marine rotifer *Brachionus plicatilis* and *B. rotundiformis* playing the most important role. However, rotifer use in the larval rearing of freshwater fish is expanding, so information on *Brachionus rubens* and *B. calyciflorus* is also included.

It is worth mentioning a growing application of rotifers in aquatic biology. A standardized acute toxicity test using rotifers hatched from cysts has been described (Snell and Persoone, 1989a and b; Snell, et al. 1991a and b) as well as a reproductive test to estimate chronic toxicity (Snell and Moffat, 1992). These tests have several advantages over existing toxicity tests because of their speed, sensitivity, reproducibility, convenience and cost-effectiveness (Snell and Janssen, 1995). Rotifers will become more widely used to assess aquatic toxicity in both marine and fresh water.

Role of Rotifers in Larval Rearing

Predator versus prey

To understand the role of rotifers in larval rearing, several biological features of feeding in larval fish and crustaceans need to be taken into account. Marine fish larvae are visual feeders that hunt during daylight (Hunter, 1980; Iwai, 1980). Reactive distance, the maximum distance that a predator can locate a specific prey, is very small for fish larvae. Beyond 0.7 to 1.0 body lengths, prey could not be detected by herring larvae and reactive distances of 0.5, 0.2, and 0.4 body lengths were recorded for plaice, pilchard, and northern anchovy larvae, respectively (Hunter, 1980). The stimuli eliciting feeding responses are poorly understood for larval fish, but in adult fish prey size is a primary characteristic (O'Brien, 1979; 1987). Movement also has an important role, probably first attracting attention to the prey. Locomotory patterns of different zooplankters determine their susceptibility to predators (Buskey, et al. 1993). Other considerations for locating prey and determining their suitability are spines, color, taste, and predator avoidance behavior (Hunter, 1980).

Initially, fish larvae are polyphagous, sighting and gulping any prey of the appropriate size. Feeding success is low during the early stages, only 6% of attempts by herring larvae are successful and 32-62% by plaice larvae (Hunter, 1980). After larvae become familiar with a prey item their performance improves (Houde and Scheker, 1980). However, if the prey is changed, larvae experience a temporary reduction in feeding success. For example, feeding success in northern anchovy larvae dropped from 80% to 40% at age 17 days when their food was changed from the rotifer *B. plicatilis* to *Artemia* nauplii. After two days on the new food, feeding success rate returned to its previous level (Hunter, 1972). These data clearly demonstrate why new feed items should be sufficiently overlapped with the previous feed items when changing diets of larval fish.

Critical prey size

Prey size dominates prey selection, the critical dimension for ingestion is prey width (Blaxter, 1965; Hunter, 1976). At the onset of feeding, most fish larvae select prey ranging from 50-100 μm (microns or micrometers) in width (Hunter 1980; Houde and Schekter 1980; Thielacker 1987). Mouth size places an upper limit on the size of prey that can be ingested and closely corresponds to the natural food eaten (Shirota, 1970). This relationship suggests that larval predators are highly adapted to consume specific prey organisms. The lower limit on prey size is probably determined by the metabolic needs of the larval predator. Adequate calories must be obtained from a feeding response for the searching, capture, handling, and ingestion behaviors of predation to yield a net caloric gain (Hunter, 1980).

As fish larvae increase in size, the size of prey they ingest also increases. Hunter (1980) plotted the relationship between prey size and larval length for 12 marine fishes. Only hake larvae, *Merluccius merluccius*, could adequately feed on *Artemia* nauplii at the onset of feeding. Hunter's data clearly show that the upper limit of prey size eaten increases faster than the lower limit. Most larval fish expand the upper size range of prey selected as their mouth size increases, but continue to actively take small prey as well.

Predator feeding success

Larvae of different species differ markedly in their predatory ability. Large differences in feeding success, capture rate, attack rates, handling times, and search rates were recorded for the bay anchovy (*Anchoa mitchilli*), sea bream (*Archosargus rhomboidalis*), and the lined sole (*Achirus lineatus*) (Houde and Schekter, 1980). As larvae grew, they became better predators with decreased search time per attack and reduced handling time for similar sized prey. Houde and Schekter argued that capture success rate at first-feeding is the critical factor determining larval fish survival and growth rates.

More studies are appearing that focus on larval feeding behavior of particular fish species. Examples include the gilthead bream *Sparus aurata* (Fernandez-Diaz, et al. 1994), weakfish *Cynoscion regalis* (Pryor and Epifanio 1993), clownfish *Amphiprion perideraion* (Coughlin, et al. 1992, Coughlin, 1993), and sunshine bass *Morone saxatilis* (Ludwig, 1994). As these data accumulate, we can expect to see more generalizations emerge about fish larval feeding behavior.

Live food utilization

Some larval fish do not have a fully functional stomach and gastric glands when they begin feeding. This makes processing of food for absorption difficult because necessary digestive enzymes are lacking (Dabrowski and Culver, 1991). Live food, like zooplankton, can be an important external source of these enzymes. When zooplankton are ingested their proteolytic (breakdown of proteins or peptides into amino acids by the action of enzymes) are released into the larval fish gut, where they contribute to the degradation of food and facilitate assimilation of nutrients. So besides providing nutrients to larval fish, live zooplankton contribute to their growth by facilitating digestion. In addition, early stage larvae prior to metamorphosis have a straight gut with limited surface area for absorption. As larvae complete metamorphosis, the gut becomes convoluted, thus increasing its absorptive capacity and slowing movement of food through the alimentary canal. The waste due to undigested food in early stage larvae is significant, therefore the initial level and quality of food must be high (Hoff, 1996).

Aquaculturists have integrated knowledge of larval fish biology into a generalized, practical protocol for rearing larvae of several marine fishes. Much of the work has been done in Europe and Japan. A good overview of a generalized pelagic marine fish larvae feeding schedule was published by Watanabe, et al. (1983a). Their scheme begins with fish larvae 2-3 mm or larger which are fed rotifers as the initial food. The rotifer *Brachionus plicatilis* is offered continuously for the first 30 days or until the larvae reach about 9.8 mm in length. When larvae reach 7 mm (about day 18), a variety of copepods are provided along with rotifers. These copepods include *Tigriopus, Paracalanus, Acartia,* or *Oithona*, most of which are obtained from natural populations in local ponds or the sea. If the copepod supply is inadequate, freshwater *Moina* or *Daphnia* are fed to the larvae. A second alternative to the marine copepods is *Artemia* nauplii which are commercially available, but add to the cost of larval rearing. Once the larvae reach 10-11 mm, they are fed minced fish, shellfish, or an artificial diet. After larvae reach 30-50 mm larval rearing is considered complete.

In some cases, when rearing pelagic species of finfish, rotifers are the sole food from first feeding (day 2-3) up to day 20. In their role as first food, rotifers are critically important for intensive rearing of most marine fish larvae and no adequate substitute has been found. In Japan, where hundreds of millions of fish, shrimp, and crab larvae are raised intensively each year, rotifers are an integral part of larviculture systems. Considering the natural ecological relationship between rotifers and fish larvae in estuaries, it is no surprise that newly hatched larvae are well adapted to capture and ingest rotifers. It is also likely that fish larvae are physiologically well adapted to utilize rotifers nutritionally. Aquaculturists worldwide have learned to exploit this natural relationship for intensive larval rearing of marine fishes. In fact, Watanabe, et al. (1983a) stated that without the mass culture of rotifers, larval rearing of marine fishes would be virtually impossible.

Species	Feeding Days	Larval Size 5 - 6 mm	Larval Size 8 - 10 mm
Chysophrys major (Red Sea bream)	20 - 30	100 - 200	100 - 2000
Lamanda yokahama (flatfish)	40 - 50	100 - 200	1000 - 2000
Amphiprion sp. (clownfish)	5 - 9	500 - 1000	not fed
Portunus trituberculatus (Japanese blue crab)	zoea to megalops	100 - 2000	1000 - 2000

Table 5.1. Number of rotifers needed per day to rear specific fish and crab larvae, (modified from Rothbard, 1975)

It has further become widely accepted opinion in Japan that the number of fish and crustacean larvae produced is limited by the amount of rotifers available (Hirata, 1979; Hirata, et al. 1983; Fukusho, 1989a; Fukusho and Hirayama, 1992). The data in Table 5.1 gives you a feeling for the number of rotifers needed per day to rear fish and crab larvae. Kafuku et al. (1989) provided a formula for determining rotifer need based on larval length or red and black sea bream. Total length of the larvae (L) and rotifer consumption (F) is expressed as: 3.934 x F = 0.0303 x L, where (L) is greater than 3.92 mm but not larger than 10.05 mm.

A feeding regime used to rear larval clownfish, *Amphiprion sp.*, is presented in Table 5.2 (Hoff, 1996). It is essential to overlap the initial live food (rotifers) with the next live food (newly hatched *Artemia* nauplii). This overlapping regime can be used to rear many fish and invertebrates. Overlapping diets allows less aggressive, slow growing individuals to survive. Abrupt changes in diets create periods of non-feeding while fish adjust to the new food. This stress often results in mortality of smaller fish and a reduction in growth. The period for utilizing rotifers and *Artemia* nauplii varies according to fish size and stage of maturity. Dry feeds (flake foods, krill meal, and salmon mash) can be sieved to various micron sizes (100-400 μm) and fed according to acceptance and size of the fish.

Type	Size	Age of Larvae In Days
		1 2 3 4 5 6 7 8 9 10 11 12 13 14 15 16 17 18 19 20-23
Rotifers	50-150μ	>>>>>>>>>>>++++++
Dry Feed	50-100μ	>>>>>>>>>>>>>+++++
Artemia	399-400μ	>>>>>>>>>>>>>>++++++
Dry Feed	100-200μ	>>>>>>>>>>>>>+++++
Krill Meal	100-200μ	>>>>>>>>>>>>>>+++++
Dry Pellets	100-600μ	>>>>>>>>>>>>>>>>>>>>>>>>>
Gelatin Mix	100-600μ	>>>>>>>>>>>

>>>>>>> = Normal feeding regime during normal growth and conditions.
++++++ = Extension beyond normal feeding range due to size variation, slow growth etc.

Table 5.2 — Modified overlapping feeding schedule for rearing larval clownfish, Amphiprion sp. at Instant Ocean Hatcheries (Hoff, 1996).

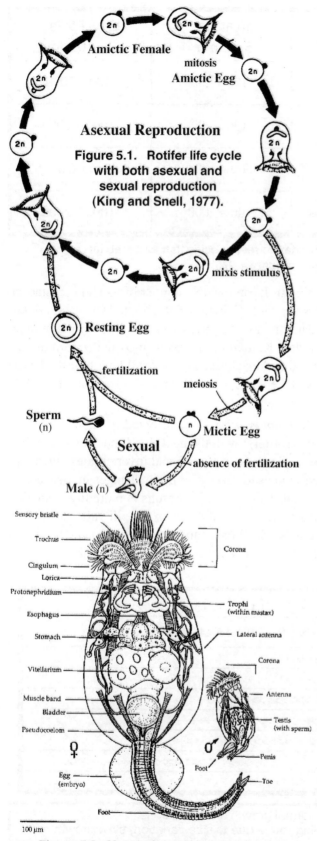

Asexual Reproduction

Figure 5.1. Rotifer life cycle with both asexual and sexual reproduction (King and Snell, 1977).

Amictic Female

mitosis
Amictic Egg

mixis stimulus

Resting Egg

fertilization

meiosis

Sperm (n)

Mictic Egg

Sexual

absence of fertilization

Male (n)

Sensory bristle
Trochus
Cingulum
Lorica
Protonephridium
Esophagus
Stomach
Vitellarium
Muscle band
Bladder
Pseudocoelom
Egg (embryo)
Foot

Corona
Trophi (within mastax)
Lateral antenna
Corona
Antenna
Testis (with sperm)
Penis
Foot
Toe

100 μm

Figure 5.2. Morphology of brachionid rotifers (Koste 1980).

In the postlarval period (day 15 to 30), live feeds usually can be stopped and only sized salmon mash and gelatin-based mix provided (Hoff, 1996). For further information regarding larval rearing and special diets refer to Hoff, 1996.

Life Cycle of Rotifers
Classification

Rotifers are in the phylum Rotifera which is a lower invertebrate phylum closely related to the round worms (nematodes). There are three classes of Rotifera: Seisionidea which are unusual marine forms, Bdelloidea which are a worm-like group that reproduce strictly asexually, and Monogononta which is the class containing *Brachionus plicatilis*, *B. calyciflorus* and *B. rubens* (Wallace and Snell, 2001). Monogononts have a cylically parthenogenetic life cycle that contains both asexual and sexual phases (Birky and Gilbert, 1971; King and Snell, 1977) (Figure 5.1). Most of the life cycle is spent in the asexual phase, but in response to certain environmental cues, sexual reproduction occurs simultaneously with asexual reproduction. The cues triggering sex are poorly understood for most species, but food, population density, the absence of physiological stress, and genetics all play a role. A female *Brachionus plicatilis* carrying asexual eggs is shown on the front cover.

Morphology

Monogonont rotifers are simple morphologically, with a body consisting of three parts: a head, trunk and foot (Figure 5.2). Locomotion and food capture is accomplished captured, by a circular band of cilia surrounding the head called a corona. Within the corona is a specialized pharynx (mastax) with hard jaws (trophi) which lead to the fluid filled stomach. When the cilia detect food the trunk contracts, pulling the mouth open around the food. If suitable, the food is moved into the mastax, crushed and flows into the stomach. Food uptake takes only seconds and is clearly visible in the stomach under a microscope. A stiff outer shell called a lorica covers and supports the body giving it a distinctive shape. The lorica sometimes has anterior and posterior spines that function as predator defense or flotation devices. Rotifers are composed of about 950 cells and have fully specialized nervous, digestive, excretory, and reproductive systems. The elongated foot at the posterior is for attachment. Since, *Brachionus plicatilis*, *B. calyciflorus*, and *B. rubens* are true planktonic rotifers, they swim continuously in the water column and rarely attach. Their planktonic behavior, slow swimming speed and lack of predator defenses make them easy prey for many predators. In fish hatcheries rotifers are considered living food capsules (biocapsules) that transmit macro and micronutrients, vitamins, and even antibiotics to fish larvae (Gatesoupe, 1982; Hoff 1996; Verpraet, et al. 1992).

Reproduction

Under good contiditions the primary mode of reproduction is asexual (parthenogenesis). . Individual brachionid rotifers do not live very long. Female life-spans at 25°C are about 6-8 days and males only live about 2 days. Despite their short lives, females have remarkable reproductive capacity. Under the optimal conditions, asexual *B. plicatilis* females begin reproducing asexually at about 18 hours old, producing asexual daughter eggs that hatch in about 12 more hours. Lifetime fecundity for a single asexual female is 20-25 daughters if food supply is adequate and water quality is good. During peak reproduction, days 1-4 of a female's life, eggs are extruded every 4-6 hr and hatch after another 12 hr. These high fecundities and short developmental times give rotifers some of the highest population growth rates recorded for animals (r = 0.7 to 1.4 offspring per female per day).

Cyst production

When rotifer population density exceeds about 70 per liter, *B. plicatilis* females begin to produce sexual daughters (Snell et 2006). Offspring from these mictic eggs hatch into small, fast swimming, nonfeeding, short-lived males. Their only purpose in their life is to find and fertilize haploid resting eggs. These fertilized eggs develop thick, tough, shells, are reddish in color and are called resting eggs or cysts which are similar to *Artemia* cysts. Rotifer cysts are encysted embryos with arrested metabolism and are capable of dormancy for several years. At 110 μm long, cysts are similar in size to asexual eggs except in color. One gram of dried rotifer cysts contains about 2 million cysts. The tough shell allows cysts to endure extreme environments, including freezing and desiccation. Because of these properties, rotifer cysts are easily dispersed by birds and other animals, sometimes ending up in places where they are unwanted.

Rotifer cysts hatch in 24-48 hours when incubated at 25°C and moderate light. The hatchlings emerge swimming and begin rapid asexual reproduction after aboutt 18 hours. As a result, cysts are excellent for starting rotifer mass cultures which can be brought up to production densities in a few days by controlling inoculation levels. This illustrates the important differences in how rotifer and *Artemia* cysts are used in aquaculture. Rotifers hatching from cysts are not used directly as food, but serve as inocula for mass cultures. Since rotifer cysts hatch into females that reproduce asexually, population size grows very quickly. In contrast, the number of *Artemia* nauplii harvested is never more than the number of cysts incubated because the nauplii are harvested long before they begin reproduction. Figure 5.3 illustrates this relationship by plotting population growth over a 10 day period. Rotifer cysts hatch and the resulting 1000 rotifers reproduce to become over 1 million at 25°C. During the same period, *Artemia* cysts hatch and begin a long series of maturation stages, but never reach reproductive maturity. Because of these fundamental biological differences, many more *Artemia* than rotifer cysts are needed by aquaculturists to provide adequate numbers of live food organisms for larval rearing.

Utilization of cysts

There are several advantages of using rotifer cysts to initiate mass cultures. Rotifer cysts eliminate the need to maintain stock cultures, resulting in considerable savings in labor. Keeping rotifers in the cyst stage rather than in stock cultures lessens the chance of contamination with ciliates or pathogenic bacteria. Rotifer cysts also can be disinfected so that the emerging rotifers are essentially bacteria-free. The time required for upscaling can be greatly reduced by using larger numbers of cysts as inocula. Figure 5.3 illustrates the effect of inoculation size on the time before a rotifer mass culture is ready for harvest. Figure 5.4 compares rotifer growth at various temperatures starting with an inoculant of 1000 cysts.

Suppose that a rotifer population size of 1 billion is required before daily harvesting of 20% of the population begins. This level of harvest would yield about 200 million rotifers per day or 600 g wet weight of rotifer biomass (1 female weighs about 3 μg wet weight). If the mass culture was inoculated with 100 rotifers, it would take 21 days before harvesting could begin. Mass cultures inoculated with 10 thousand, 1 million, and 100 million rotifers would be harvested in 15, 9, and 5 days, respectively. Shorter upscaling time for rotifer cultures makes it easier to match food production to the feeding demands of hungry larvae. Better larval nutrition will almost certainly lead to increased survival and growth rates. Rotifer cysts have shelf life of over one year at room temperature. However, for maximum survival it is recommended that cysts are stored in dry form and in the freezer. Thus providing insurance for aquaculturists to rapidly establish a culture when necessary. Rotifer cysts for both the marine species *Brachionus plicatilis* and the freshwater species *Brachionus calyciflorus* are available from Florida Aqua Farms under the trade name Resting Rotifers™.

Reducing upscaling time using cysts as inocula also has application in disease control. If an infection spreads through a hatchery, many aquaculturists halt production, dismantle the rotifer and larval cultures, and treat the entire facility with disinfectant. Hatchery managers hesitate to use this treatment because the time lost in restarting the rotifer cultures reduces production. Hatching a large batch of rotifer cysts allows a quick return to production, minimizing down time for treatment.

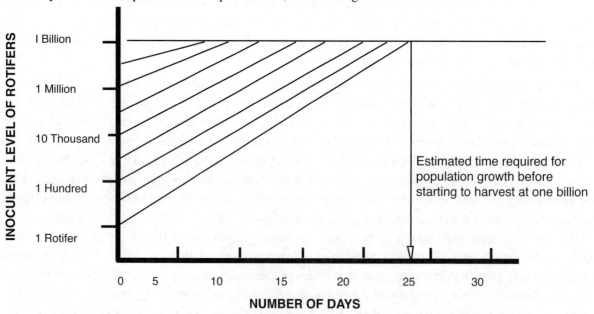

Figure 5.3. Inoculation size and time required to reach a rotifer population of one billion.

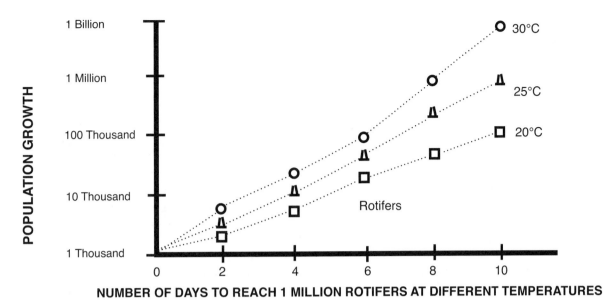

Figure 5.4 — Comparison of rotifer growth over a ten day culture period at various temperatures starting with one thousand rotifer inoculant per temperature.

Several papers have discussed other aspects of utilizing rotifer cysts in aquaculture (Lubzens, 1981; Snell and Hoff, 1985; Lubzens, 1987; Snell and Hoff, 1988; Hagiwara and Hino, 1988; Lubzens, et al. 1989; Hagiwara, et al. 1989, Hagiwara and Hirayama, 1993, Hagiwara, 1994) and the general biology of rotifer cysts (Pourriot and Snell, 1983). Hagiwara, et al. 1995 showed that cysts hatched better in presence of light, 44% compared to 0.5% in the dark. Hatching in the dark could be enhanced by treating cysts with hydrogen peroxide or prostaglandins yet the presence of light consistently yielded better results. Recent work has shown that 250-310 nm light is best for hatching *B. plicatilis* cysts

Differences Among Rotifer Strains

Strains

Brachionus plicatilis, *B. rotundiformis*, *B. calyciflorus*, and *B. rubens* are found in a variety of different habitats all over the world. It is therefore not surprising that considerable variation exists between strains from different locations. Snell and Carrillo (1984) reported on differences in size among 13 strains of *B. plicatilis* which ranged in length from 123-292 μm and in width from 114-199 μm. In comparison, the smallest newly hatched *Artemia* nauplii are 400 μm long and 250 μm wide. Extensive work on characterizing the growth requirements of several *B. plicatilis* strains useful in aquaculture has been done in Japan (Fukusho, 1989a; Hirayama, 1990). Larger forms of *B. plicatilis* have been reported from remote areas of Australia which range up to 440 μm long (Koste, 1980). Body size of brachionid rotifers is primarily determined genetically however, environmental factors like temperature, salinity, or food type can alter body size by about 15% (Serra and Miracle, 1983; Snell and Carrillo, 1984). For example, rotifers fed on baker's yeast are usually larger than those fed on live algae.

Different sized *B. calyciflorus* strains have been characterized by Rico-Martinez and Dodson (1992) They compared lorica lengths and widths of 4 strains collected from Florida and Wisconsin. Lengths ranged from 195 to 277 μm and widths from 121 to 164 μm. As in *B. plicatilis*, these size differences among *B. calyciflorus* strains provide aquaculturists with the opportunity to customize prey size to the specific needs of the larval predator.

L and S strain rotifers

Brachionus plicatilis and *B. calyciflorus* are found world wide. However, populations from different geographical regions have quite different morphological, physiological, and behavioral characteristics. This variation among strains has been documented in *B. plicatilis* by several papers (Fukusho and Okauchi, 1982, Yufera, 1982; Serra and Miracle, 1983; Snell and Carrillo, 1984; Snell and Winkler, 1984) and has been categorized into two major groups by Fukusho (1983). These have been called "S" and "L" types which differ in their size, shape, anterior spines, and temperature optima. Hirayama (1990) has compared several strains of S and L types for these characteristics and reported that S type strains tend to be round, ranging in length from 150-220 μm as compared to 200-360 μm for oblong L types. The optimum temperature for growth of S types is 28°-35°C and that of L types is 18°-25°C. Genetic characterization of these strains has led to the conclusion that S and L types are different species (Fu, et al. 1991a and b). Recognition of these differences has led to the naming of L type as *Brachionus plicatilis* and S type as *B. rotundiformis* (Segers, 1995). Molecular genetic analyses are being applied to *Brachionus* strains used in aquaculture (Boehm et al. 2000). Additional work tohas continued to uncover new species (Gomez et al. 2002, Derry et al. 2003, Suatoni et al. 2006), suggesting that the B. plicatilis species complex is comprised of more than 15 species. Two of these have been formally named: B. ibericus (Ciros-Perez et al. 2004) and B. manjavacas (Fontaneto et al. 2007). Application of a variety of molecular techniques to classify rotifers used in European hatcheries (Papakostas et al. 2006) has shown the predominance of the Brachionus sp. Cayman biotype. Other biotypes present at lower frequencies include B. plicatilis sensu strictu, Brachionus sp. Nevada, and Brachionus sp. Austria, all L-type species. The Brachionus sp. Cayman biotype outcompetes other Brachionus species in the high food conditions that are common in most hatcheries (Kostopoulou and Vadstein 2007). Hatcheries that trade rotifer cultures based solely on morphology may actually be using different species with different temperature and salinity optima.

Utilization of S strain rotifers

This variation in rotifer body size provides aquaculturists with the opportunity to match larval food size to predator mouth size. The larger species *B. plicatilis* would be more appropriate for fish with large mouth sizes because they contain more nutrients per rotifer. If these were too large for certain larval fish to take at first feeding, as might be the case for fish like grouper, the smaller *B. rotundiformis* species may yield better initial results. Most species of fish larvae can be reared on *B. plicatilis*. Our recommendation is to begin with *B. plicatilis* and if there are problems that are suspected to be related to food size, try *B. rotundiformis*. Another consideration would be the ambient temperature of rotifer cultures, if it routinely exceeds 28°C, culturing of *B. rotundiformis* may yield better results.

An illustration of how different size rotifer strains might be used in larval culture has been provided by Polo, et al. (1992). They fed *Sparus aurata* larvae a mixed diet of *B. rotundiformis* and *B. plicatilis*. Larvae less than 8 days old (larval length < 4.3 mm) selected the smaller rotifers and after day 13 (larval length > 5.1 mm) the larger rotifer was chosen. The presence of the smaller rotifers enabled more larvae to begin feeding (55-80%) than when small rotifers were absent (15-45%). Greater growth rates and survival were obtained on the mixed diet than on *B. plicatilis* alone. However, total larval length is not as important as mouth size in determining what size rotifer to use. Hagiwara et al. (2007) described ways to manipulate rotifer culture conditions to produce rotifers with particular size, population growth and stability characteristics. They also reviewed genetic manipulations including crosses and molecular approaches to match rotifer traits to the needs of larvae.

Developing S cultures from existing L cultures

Any population of any zooplankton follows a typical bell shaped curve where most individuals cluster around some mean size with smaller and larger individuals on either end of the curve. You could enrich for smaller individuals if you selectively sieved smaller individuals from your cultures. Initially the majority of the smaller individuals would be newly hatched juveniles yet, if you consistently sieve smaller individuals and set up new cultures from these populations then you could eventually genetically modify the population.

Most newly hatched fish larvae only need small rotifers for 1-3 days before they can accept full size *B. plicatilis* rotifers. If you maintained large cultures of rotifers and sieved for only the smaller individuals for first few days you would not need to maintain cultures of two different rotifer species.

In addition, as mentioned in the previous section above, populations can vary in size by about 15% due to environmental conditions including temperature, salinity, or food type so smaller sizes cann be produced by manipulating these environmental factors.

Utilization of other rotifer species

The main rotifer species currently utilized in marine aquaculture are Brachionus plicatilis and B. rotundiformis and B. calyciflorus in freshwater aquaculture. There are, however, more than 2000 species of rotifers (Wallace and Snell 2001, Segers 2008), the vast majority of which never have been investigated for use in aquaculture. Knowledge is accumulating on other rotifers that may also be suitable for aquaculture applications. A prime example is Synchaeta which grows in the right temperature and salinity range (Oltra and Todoli 1997; Bosque et al. 2001) and has a good fatty acid profile (Oltra et al. 2000). Another example is Colurella decentra (Suchar and Chigbu 2006), an especially small rotifer (49-93 μm) found in many coastal marine habitats. It is easily cultured on a variety of marine microalgae typically cultured in hatcheries. More innovative aquaculturists may want to explore rotifer biodiversity more thoroughly than it has been to date.

Physical and Chemical Requirements - Marine Rotifers

Brachionus plicatilis is found in many different environments from seawater to salt lakes and has a wide range of physiological tolerances (Walker, 1981; Serra and Miracle, 1983). Strains found throughout the world vary greatly in their tendency towards cyst production, but for most applications, conditions are chosen that allow asexual reproduction to dominate. The objective in aquaculture is to promote high rates of asexual reproduction and minimize sexual reproduction. A summary of the physical and chemical requirements for several aquacultured strains can be found in Fulks and Main (1991).

Salinity

Salinity has a marked effect on rotifer production in mass cultures. *Brachionus plicatilis* tolerates salinities from 1-97 ppt. however reproductions can only take place at salinities below 35 ppt (Lubzens, 1987). Best growth occurs at intermediate brackish water salinities between 10-20 ppt (1.007-1.014 density). Abrupt salinity changes of 10 ppt (1.0069 density) or more are likely to kill *B. plicatilis*, but gradual acclimation to higher salinity greatly reduces mortality. Keeping culture salinities close to the larval rering levels will help reduce osmotic stress when the rotifers are transfered. Maximum reproductive rates in an Israeli strain were recorded at 4-10 ppt (1.0023-1.0069 density) and then declined linearly with increasing salinities (Minkoff, et al. 1985). At 35 ppt (1.0260 density) reproductive rates were less than the maximum, and at 48 ppt (1.0351 density) they were only about 1/10 the maximum. This suggests that some strains of *B. plicatilis* are actually brackish water rather than truly marine rotifers. *B. plicatilis* also tolerates saline waters with a variety of ionic compositions as is found in inland salt lakes (Walker, 1981). However, the best growth is obtained when the ionic composition of the medium closely approximates seawater. Salinities higher than 35 ppt (1.0260 density) markedly suppress sexual reproduction, called mixis in rotifers (Lubzens, et al. 1985, Snell and Hoff, 1985). Suppression of sexual reproduction (mixis) is desirable in mass cultures because the presence of males and cysts reduce population growth rate and overall rotifer yield (Snell, 1987). Some strains do not produce males and cysts even at lower salinities and therefore are particularly well suited for mass culture (Lubzens, et al. 1984).

Temperature and light

Temperature strongly influences reproductive rate, and each rotifer strain has its own particular optimum (Hirayama, 1990). Rotifer species also differ in their optimal temperature. *B. rotundiformis* has an optimal growth at 28-35°C (82-95°F) whereas, *B. plicatilis* reaches optimal growth at 18-25°C (64-77°F). In general, temperatures should be maintained between 20-30°°C (68-86°F) (Hirayama and Kusano, 1972). Rotifers survive higher and lower temperatures but, low temperatures result in lower production and high temperatures cause thermal stress and depress production. Different strains vary in their optimum temperature for growth (Snell and Carrillo, 1984); some thermophilic strains have been identified with maximum yields at 30°-35°C (Pascual and Yufera, 1983). However, mass culture stability is easier to maintain at intermediate temperatures.

Several other environmental factors influence rotifer production and the following are general guidelines. Particular strains, however, may have requirements that differ considerably from these values. Light intensity should be between 2000-5000 lux (200-500 foot-candles) and provided on a photoperiod of 16 hours light, 8 hours dark.

PH and Aeration

Moderate to low aeration should be provided (filtration is not necessary) and pH should be maintained between 6.5. and 8.0. Free ammonia concentrations in rotifer cultures should not exceed 1 mg/liter (ppm). Because free ammonia concentrations are lower at low pHs, some aquaculturists keep pH low (6.0 to 6.5) in their mass cultures (Yoshimura, et al. 1995). We feel that a pH of 6.9 to 7.8 more is easier to maintain in mass cultures and enhances their stability. Furukawa and Hidaca (1973) observed highest densities at a pH of 7.3 to 7.8. Low culture pH allows culture densities to increase about 10-fold over typical seawater pH of 8.2. Free ammonia concentration remained at 2 ppm despite very high population densities of up to 34,000 *B. rotundiformis* per ml.

Physical and Chemical Requirements - Freshwater Rotifers

The rotifers *Brachionus calyciflorus* and *B. rubens* have become the most commonly cultured rotifers in freshwater aquaculture. Several studies have examined rotifer performance in freshwater mass cultures in an attempt to identify optimal culture conditions (Schluter, 1980; Schluter and Groenweg, 1981; Groenweg and Schluter, 1981).

Temperature and light

Critical experiments to define optimal temperature have not been done, but several observations suggest that these rotifers tolerate high temperatures. Groenweg and Schluter (1981) observed *B. rubens* populations in outdoor algal production ponds used to treat piggery wastes from June-August. The minimum and maximum temperatures were 10° and 25°C, (50° and 77°F) respectively. Rotifer production remained high throughout this period with rotifer densities fluctuating between 200-600 /ml. Our experience with *B. calyciflorus* and *B. rubens* mass cultures suggests that our strains grow well 15-31°C (59-88°F), but optimum temperatures are not known. More work is being done on the culture of B. calyciflorus under tropical conditions (Arimoro 2006).

Water composition

Ionic composition of freshwater varies greatly from site to site. We have had good results with *B. calyciflorus* mass cultures in moderately hard well water and in a defined, synthetic medium consisting of 96 mg $NaHCO_3$ (sodium bicarbonate), 60 mg $CaSO_4$ $2H_2O$ (dihydrous calcium sulfate), 60 mg $MgSO_4$ (magnesium sulfate), and 4 mg KCl (potassium chloride), in one liter of deionized water (mg/L = ppm). Hardness of this medium is 80-100 mg $CaCO_3$ /L (calcium carbonate) and alkalinity 60-70 mg/L. *B. calyciflorus* tolerates a wide variety of freshwater, but care should be taken to confirm the suitability of the local freshwater before setting up large mass cultures.

B. calyciflorus tolerates a wide variety of freshwaters, but care should be taken to confirm the suitability of the local freshwater before setting up large mass cultures.

Other conditions for successful mass culture of *B. calyciflorus* include the following. Optimal pH in mass cultures is 6-8 at 25°C (77°F), with upper and lower lethal limits of 10.5 and 3.5, respectively (Mitchell, 1992). Oxygen levels of at least 1.2 mg/L (ppm) sustain good growth, but O_2 levels below 0.7 mg/L are strongly limiting. The reproduction of *B. calyciflorus* in mass cultures is inhibited by free ammonia levels of 3-5 mg/L (ppm), with mortality occurring at only slightly higher concentrations.

Food Requirements - Marine Rotifers

Rotifer weight

The dry weight of a single rotifer depends upon its nutritional state and body size, but ranges from 0.12 to 0.36 μg (micrograms) per female, excluding eggs (Doohan, 1973, Theilacker and Kimball, 1984, Lubzens, et al. 1989). During extended periods of rapid growth and abundant food, single female dry weight can reach up to 0.62 μg. The ash-free caloric content of rotifer biomass ranges from 4.8 to 6.7 cal /mg (Lubzens, et al. 1989). In the absence of food, rotifers lose 18-26% of their body weight per day at 25°C (77°F). This rapid weight loss during starvation is the result of a high metabolic rate. It is therefore important to avoid even 24 hr lapses in rotifer feeding, because the nutritional quality and caloric content of rotifers deteriorates quickly (Szyper, 1989).

Live microalgae feeds

Food quantity and quality play an important role in the growth of rotifers. *B. plicatilis* and *B. rotundiformis* eat a wide variety of microalgae species, but the marine alga *Nannochloropsis*, is considered one of the best. This species was introduced to aquaculture as "marine Chlorella" by the Japanese, but it is now recognized as *Nannochloropsis* (Maruyama, et al. 1986). Under optimal conditions, a single rotifer can consume up to 200 *Nannochloropsis* cells per minute. Rotifer feeding rates change with environmental conditions like temperature, salinity, as well as food quantity and type (Hirayama and Ogawa, 1972). *Nannochloropsis* concentrations of 2 million cells/ml were sufficient to maintain maximum ingestion rates. When rotifer densities are high (>100 /ml), almost all of the algae is consumed within one hour, even at an initial *Nannochloropsis* density of 2 million cells/ml. Consequently, food must be supplied at much higher concentrations or, preferably, at frequent intervals. One of the most difficult things in aquaculture is maintaining a proper balance between adequate amounts of food and degradation of water quality from over feeding.

At Instant Ocean Hatcheries, about 378 liters (100 gallons) of *Pyramimonas* at a density of 2 million cells /ml was used to produce 20 to 50 million marine rotifers (average 33 million) over a five day period (approximately 250,000 algae cells per rotifer). Experimentation at Florida Aqua Farms on freshwater rotifers found that it took 95 liters (25 gallons) of *Nanochloris* (a freshwater green alga) at 5.5 million cells/ml to produce 36 million rotifers in five days (about 145,000 algae cells per rotifer). Reduced feeding of algae was possible by the addition of a yeast-based, fortified liquid food (Roti-Rich™, Florida Aqua Farms). Recent experimentation at this facility has shown that further demand for live algae can be greatly reduced and still obtain adequate rotifer densities. Using a combination of 20 to 50 ml live microalgae and 2 ml of Roti-Rich™ per 4 liters per day can yield rotifer densities of 161 to 234/ml in just 5 to 8 days starting with inoculant levels of 36 to 66 rotifers/ml. It is likely that microalgae stimulates higher egg production by providing a rich diversity of nutrients while yeast feeds provides for the bulk caloric needs of the populations which yeilds lower level egg production.

Other microalgae that have been used for marine rotifer culture include: *Nanochloris, Chaetoceros, Dunaliella, Pyramimonas, Isochrysis,* and *Tetraselmis*. However, *Nannochloropis, Isochrysis,* and *Tetraselmis* are prefered by most aquaculturists for their nutritional value. Okauchi and Fukusho (1983a) presented evidence that *Tetraselmis tetrathele*, a motile microalgae, may even be better than

Nannochloropsis as a marine rotifer food. Like *Nannochloropsis*, *Tetraselmis* can be easily mass cultured (Okauchi and Fukusho, 1983b; Trotta, 1983; Camacho, et al. 1990). The large size (9-10 μm wide x 12-14 μm long) and the favorable nutritional characteristics it transfers to rotifers make *Tetraselmis* an excellent primary or secondary food. Our experience at Florida Aqua Farms suggests that routine small additions of *Tetraselmis* to a primary diet of *Nannochloropsis* and yeast or Roti-Rich™ enhances rotifer reproductive rates. There are some data to suggest that *Isochrysis galbana* fed rotifers have superior nutritional qualities for certain larval fish. Using a combination of microalgae is often advantageous when working out a diet for new larval species.

Dry microalgae feeds

Recent advancements in microalgae mass culture and biomass preservation have introduced possible new options for rotifer culture (Snell, 1991). Spray drying of algal biomass has proven effective in commercial production (Barclay and Zeller, 1996). This technique yields a dried product that retains much of the nutritional quality of live cells (Biedenbach, et al. 1990; Snell, et al. 1990). An example of how dried algae can be used for rotifer culture was provided by Snell, et al. (1990). The alga *Nannochloropsis salina* was cultured by Earthrise Farms (Calipatria, California), the biomass harvested, and preserved by various methods. The effectiveness of this algal product to support rotifer population growth was tested using a standardized population growth bioassay with *B. plicatilis*. Algal biomass was preserved using three types of freezing, saline, and two types of spray drying. Rotifer population growth rates (r) on the frozen and saline preserved algae were only 35% that of live cells. Spray dried *Nannochloropsis* yielded growth rates about 70% those of live cells, despite the fact that the spray drying process was not fully optimized. Although the results look promising, the cost of producing dried algal products remain high.

Side-by-side comparisons were made between live *Nannochloropsis* cells and spray-dried *Nannochloropsis*, dried *Tetraselmis suecica* (Cell Systems Ltd., Cambridge, England), Microfeast L-10 yeast (Provesta, Barttlesville, Oklahoma), Culture Selco (Artemia Systems, Belgium), and 7B yeast (Fleischmann, New York). Live *Nannochloropsis* was provided at 5 X 10^6 cells per ml, dried *Nannochloropsis* and *Tetraselmis* at 100 μg per ml, and yeast at 80 μg per ml. Both dried algal products yielded better rotifer growth than the yeasts, but only about 70% as high as live *Nannochloropsis* cells. Therefore, with current production techniques, rotifer diets of exclusively dried algae cannot match the biomass yield of live algae. Dried algae, however, could replace 80-90% of live cells without productivity loss (Snell, et al. 1990). It is likely that the quality of dried algae will continue to improve over the next several years and hopefully the costs to produce these products will decline.

The ability of freeze-dried algae to support rotifer population growth was tested by Yufera and Navarro (1995). They fed freeze-dried cells of *Nannochloropsis oculata*, *N. gaditana*, *Nannochloris oculata*, and *Tetraselmis suecica* to *Brachionus plicatilis* and *B. rotundiformis*. For both rotifer species, the best population growth was on a diet of freeze-dried *Nannochloropsis oculata* and the poorest on *Nanochloris oculata*. These experiments demonstrate the potential of dried algal biomass to support rotifer growth. More work is necessary, however, before optimal culture and drying conditions are identified for producing algal biomass of the highest nutritional quality.

Microalgae frozen and paste feeds

Microalgal biomass has been successfully preserved as rotifer feed using refrigeration (-1 to +3°C) four up to 3 months or freezing at -20°C (2+ years). Frozen *Nannochloropsis* was used to replace live algae by Hamada and Hagiwara (1993) for rotifer cyst production. Lubzens, et al. (1995) compared frozen *Nannochloropsis* sp. with live and dried algae, as well as yeast. Reproductive rate of *B. plicatilis* on a diet of frozen *Nannochloropsis* was 81% of that on live *Nannochloropsis* cells. Rotifer reproductive rates on diets of dried *Nannochloropsis*, live yeast, and dried yeast were 58%, 76% and 38% of the rate on live *Nannochloropsis* cells respectively.

Refrigerated, concentrated *Nannochloropsis* paste can be stored up to three months with minimal loss of nutritional quality. If retained longer it is recommended to freeze the paste into ice cubes and then store in a good plastic bag. This makes it easier for aquaculturists to match algae production with rotifer feeding needs. Frozen *Nannochloropsis* cells will support good rotifer population growth. Lubzens, et al. (1995) also demonstrated that the fatty acid compositions of rotifers fed frozen *Nannochloropsis* verses live *Nannochloropsis* were very similar. This indicates that rotifers fed frozen *Nannochloropsis* can be used as larval fish feed without further enrichment.

Cryopresevation of *Nannochloropsis, Tetraselmis, Isochrysis,* and *Chaetoceros* are now commercially available as paste in the USA. Addition of cryopreservatives prevents the algae from freezing even at temperatures below -20°C (-68°F). This allows a shelf life of up to one+ years for some species. However, gradual (2-5%) loss of nutritional value after one year may occur. Convenience and effectiveness of algae paste for commercial operations has caused concern in rotifer culture with regard to the use of cryopreservatives and there long term effects with regards to nutritional value.

Yeast as a feed

In addition to microalgae, rotifers can be raised on yeast. Baker's yeast has a small particle size (5-7μ diameter), high protein (45-52%) content and low cost. Yeast has been used very successfully in a number culture systems (Lubzens, 1987). However, problems often arise when feeding yeast the sole diet, due to its nutritional deficiencies. Rotifers fed yeast only show a slow growth rate and low fecundity leading to an extended culture period to achieve desired densities. On a yeast diet, sudden unexpected density decreases may occur due to deterioration of water quality (Komis 1992). Under certain conditions unenriched, yeast fed, rotifers are often an inadequate food source for marine fish larvae, particularly in terms of survival, normal development (swim bladder inflation) and optimum growth rate (Witt et al., 1984; Dendrinos and Thorpe, 1987; Gatesoupe and Le Milinaire 1984). Fresh, active baker's yeast is normally fed at 1-3g/10 million rotifers and inactivated dry yeast at 0.3-0.9g/10 million rotifers. Cell densities of liquid active yeast vary from 1.10x10 to 2.40x10/ml . Many types of yeast are suitable for rotifer culture however, other considerations such as contamination have to be considered. For example, brewer's yeast often contains higher levels of copper and other metals. Activated baker's yeast is good, yet it is difficult to administer to cultures on a continuous basis without causing excessive bacterial growth and rapid deterioration of water quality. Overfeeding is consistently a problem with yeast and yeast-based products. Indications of over feeding are excessive slime on the sides of the culture vessel, foul odor, or long strings of flocculent particles suspended in the water.

Providing yeast in a 1:1 ratio with *Nannochloropsis* yielded rotifer reproductive rates equal to those on *Nannochloropsis* alone (Hirayama and Watanabe, 1973). Yeast can therefore replace at least a portion of the *Nannochloropsis* required by rotifer mass cultures, resulting in considerable savings. Besides baker's yeast, several species of marine yeasts have been developed for rotifer culture (Furukawa and Hidaka, 1973, James, et al. 1987). Although yeast is very effective for rotifer growth, yeast-fed rotifers are not nutritionally sufficient to rear most marine fish larvae (Watanabe, et al. 1983a).

Nutritional quality of yeast-fed rotifers can be improved by enrichment with live microalgae or omega-3 fatty acid emulsions prior to feeding larval fish. Descriptions of these enrichment techniques are discussed in Rotifer Enrichment. In addition, the reproductive rate of yeast-fed rotifers is lower than algae or a yeast/algae mixture.

Nutritional quality of yeast can be improved with vitamin supplements (Hirayama and Satuito, 1991). *Brachionus plicatilis* has a dietary requirement for vitamin B_{12} (Scott, 1981; Hoff, 1996) that often is not met in yeast-fed mass cultures. When yeast-fed batch cultures were supplemented with 1.4 μg B_{12}/ml, they exhibited a 70% increase in reproductive rate (r) (Hirayama and Satuito, 1991). Likewise, supplementation with 3 μg vitamin A, 0.1 μg vitamin D, 0.5 μg vitamin E, or 0.4 μg vitamin C per ml caused significant increases in reproductive rate and net fecundity per female. Squid liver oil improved rotifer growth when added at 4 μg/ml to mass cultures, presumably by supplying essential fatty acids only available in limited quantities in yeast. This suggests that the nutritional quality of yeast for rotifer culture can be substantially improved with supplements of vitamins and fatty acids.

Yeast Based Commercial Feeds

Florida Aqua Farms, (FAF), INVE, Aquafauna and others produce convenient, fortified yeast-based products. Roti-Rich™ made by FAF contains particular amino acids, essential fatty acids, vitamins and minerals that promote rotifer growth and nutritional balance. Experimentation using this formula as a sole food demonstrated that rotifer cultures with densities of 100 to 300/ml can be maintained for more than two years. INVE has produced several yeast based products including; Culture Selco™ and Protein Selco™. All of these formulas are based on use of deactivated yeast cells.

Food Requirements - Freshwater Rotifers

Live microalgae feeds

After testing several microalgae species as diets for *B. calyciflorus* and *B. rubens*, our experience has been that *Nannochloris, Nannochloropsis,* and *Chlorella* work well. However the ease of culture and high nutritive quality for *B. calyciflorus* and *B. rubens* makes *Nannochloropsis* and *Chlorella* the alga of choice for many aquaculture applications. Specific environmental conditions at some sites or the fish specie being cultured may require other algal species to be fed to the rotifers based on their environmental tolerance or specific nutrient and chemical values. Several other freshwater algal species also may be suitable. Of the eight algae investigated by Schluter (1980), *Scendesmus costato-granulatus, Kirchneriella contorta,* and *Chlorella fusca* were found to be best. Pourriot, et al. (1987) found *Chlorella pyrenoidosa, Phacus pyrum* and *Ankistrodesmus convolutus* equally suitable as foods for *B. rubens*. It therefore appears that many kinds of microalgae are suitable for freshwater rotifer culture. Pilarska (1972) compared *Chlorella vulgaris* with the bacterium *Aerobacter aerogenes* and found that algae-fed rotifers produced about twice as many eggs as bacteria-fed rotifers. From this work and that of Yasuda and Taga (1980) and Ushiro, et al. (1980) it can be concluded that bacteria alone are not as good as microalgae for rotifer feeds. Although bacteria growing on decomposing yeast cells are an important source of rotifer nutrients (Hirayama and Funamoto, 1983).

The feeding rate of *B. rubens* is somewhat higher than that of *B. plicatilis* on a *Chlorella* diet. At a *Chlorella* concentration of 3 million cells/ml, a single *B. rubens* female consumed about 700 cells/hr at 20°C (Pilarska, 1977). Feeding rate increased nearly linearly from ten thousand up to 5 million cells/ml, but at higher cell densities no further increases in ingestion rate were recorded. Filtering rate at ten thousand cells/ml is maximal at about 12 μl per female per hr and declines logarithmically at higher cell densities. These data illustrate the voracity of rotifers and the difficulty for aquaculturists of supplying mass cultures with adequate amounts of food. Yet, little data is available to substantiate whether higher consumption of algae produces higher densities of rotifers or if they are more nutritious.

Yeast Based Feeds

As is the case for the marine rotifer *B. plicatilis*, yeast is a good food supplement for freshwater *B. calyciflorus* and *B. rubens* cultures. All of the cautions described in the *B. plicatilis* section apply to freshwater rotifers and should be underscored because their effects in freshwater are usually magnified. We have used Roti-Rich™ successfully as a sole food in culturing *B. calyciflorus* for several months, but have not achieved population densities as high those with *B. plicatilis*. Although a yeast diet is a simple means of rotifer culture, algae-fed rotifers are always more nutritious for larval fish than yeast-fed rotifers. Consequently, if freshwater rotifers are cultured on yeast or yeast based products alone, enrichment with an omega-3 fatty acid emulsion (Watanabe, et al. 1983a) may be necessary to achieve maximum nutritional value. It also has been demonstrated that feeding a suitable microalgae the day before and an hour or so prior to utilizing the rotifers will also sufficiently enhance their fatty acid profile.

Bacteria in Rotifer Cultures

Rotifer mass cultures are complex microcosms of bacteria, algae, protozoans, and rotifers. Recent work has provided insight into the role of bacteria in supporting rotifer growth and maintaining mass culture stability (Hino, 1993). Substantial quantities of bacteria have been found in rotifer mass cultures. Muroga, et al. (1987) reported 2-20 million colony forming units (CFUs) on rotifers, depending on the type of nutrient agar used to grow the bacteria. The species composition was 56% *Pseudomonas*, 14% *Moraxella*, 12 % *Vibrio*, and 18% other species. Miyakawa and Muroga (1988) found 1-100 million CFUs on rotifers and 1000 to 1 million CFUs/ml in the culture water. These and the results of Yu, et al. (1989) demonstrate that the density of bacteria in rotifer mass cultures is typically high.

Techniques are available to manipulate bacterial populations in rotifer mass cultures to increase production and reduce unwanted species. Probiotic bacteria like *Lactobacillus plantarum* are added to competitively reduce the explosive growth of unwanted bacteria such a *Vibrio* which may occur after enrichment. Enrichment of cultures generally induces a shift in the bacteria composition from *Cytophaga/Flavobacterium* dominance to *Pseudomonas/Alcaligenes* dominance. This change is partly due to a bloom of fast growing opportunistic bacteria, favored by high substrate levels (Skjermo and Vadstein, 1993). Hirata et al. (1998) pre-cultured some 40 species of probiotic bacteria and then added 20 ml per liter to rotifer culture water and observed enhanced rotifer population growth. Makridis et al. (2000) bioencapsulated probiotic bacteria into rotifers before inoculating mass cultures. These probiotic bacteria remained present in the rotifers for up to 24 hours, replacing the potentially harmful opportunistic bacteria. Addition of terrestrial lactic acid bacteria to rotifer cultures increased mass culture population growth rate 8-13 times over controls (Planas et al. 2004). Roles and use of probiotic bacteria in rotifer culture has been reviewed by Verschuere et al. (2000). Importance of bacteria as biological control agents in rotifer cultures is likely to grow.

Bacteria also have been utilized in rotifer mass cultures as food, producers of vitamin B_{12}, and producers of eicosapentanoic acid (EPA). A variety of bacterial species have been investigated as food for rotifer mass cultures (Ushiro, 1980; Yasuda and Taga, 1980). Rotifer yields on bacterial diets are not comparable to those of algae fed cultures. Some bacteria, however, when added to algal diets, can improve yeilds substantially. Douillet (2000) found that the marine bacterium *Alteromonas sp.* as a sole diet supported good population growth of B. plicatilis, with r values of 0.6-0.8. These are about 2/3 of the population growth rate on algal diets.

Bacteria in mass cultures contribute significantly to meeting the vitamin B_{12} requirement of *B. plicatilis*. The B_{12} demand of *B. plicatilis* is 1-1.5 pg/rotifer, depending on the strain (Yu, et al. 1988). An average *B. plicatilis* contains 1-2.7 pg B_{12} (Yu, et al. 1989). The major B_{12} producing bacteria typically found in rotifer cultures are *Pseudomonas* (66%), *Moraxella* (11%), *Bacillus* (9%), *Vibrio* (9%), and others (4%). Addition of these bacteria at a concentration of 100 million to 100 billion CFUs/ml can improve rotifer growth dramatically.

Some bacteria produce eicosapentanoic acid (EPA) and one strain, *Shewanella putrefaciens*, has been used to improve the growth and nutritional quality of *B. plicatilis* (Watanabe, et al. 1992). Other bacteria, like certain strains of *Vibrio alginolyticus*, produce an exotoxin that can kill *B. plicatilis* (Yu, et al. 1990). This exotoxin may be responsible for the sudden mass mortality that is sometimes observed in rotifer cultures.

Lactic acid bacteria (LAB) like *Lactobacillus* and *Carnobacterium* have been isolated from *B. plicatilis* (Gatesoupe, 1991). When LAB were cultivated and added daily to the rotifer enrichment medium, larval turbot were more resistant to pathogenic *Vibrio* (Gatesoupe, 1994). Introduction of specialized bacterial strains like LAB into rotifers may be more effective than antibiotics in regulating larval fish pathogens.

We have found that when rotifer cultures are transferred, adding about 10%-20% of the old culture water increases initial survival. This technique inoculates beneficial bacteria and reduces the lag time in establishing new cultures.

Nutritional Value of Rotifers

Biochemical composition

Biochemical composition of rotifers and therefore their nutritional value to larval fish is determined by diet. An extensive analysis of the biochemical composition of *B. plicatilis* was completed by Watanabe, et al. (1983a). *Nannochloropsis* fed rotifers were composed of 75% protein, 22% lipid and 3% ash by dry weight, whereas the same values for yeast fed rotifers were 71%, 17%, and 12%, respectively. The content of several trace metals and 18 common amino acids in rotifer biomass was also reported. The digestibility of rotifer protein was 89-94% regardless whether the diet was yeast or *Nannochloropsis*. By comparison, the digestibility of *Artemia* protein was only 83% for carp and 89% for rainbow trout. The overall conclusion of this study was that live zooplankton are valuable food sources for larval fish because of their amino acid composition and high digestibility. Other studies have also examined the biochemical composition of rotifers fed different diets (Scott and Baynes, 1978; Ben-Amotz, et al. 1987, Frolov, et al. 1991, Nagata and Whyte, 1992, Tamara, et al. 1993, Fernandez-Reiriz and Labarta,1996).

When yeast became popular in Japan as a replacement for algae in rotifer mass cultures, problems began to arise with larval mortality. Sudden heavy larval losses were experienced by many hatcheries when they used exclusively yeast fed rotifers (Kitajima, et al. 1979; Fujita, 1979). It was soon discovered that the high larval mortality could be avoided by culturing rotifers on both yeast and *Nannochloropsis* or or by feeding rotifers *Nannochloropsis* for a few days just before feeding them to larval fish. Clearly, something in *Nannochloropsis* was able to compensate for the nutritional inadequacy of yeast. After extensive research, the essential fatty acid content of rotifers was identified as the critical component influencing larval survival (Watanabe, 1993).

Essential fatty acids

The essential fatty acid requirements of fish differ markedly from species to species, but there are some generalizations. It appears that highly unsaturated fatty acids (HUFA) of the 20:5w3 and 22:6w3 type are dietary requirements for many marine fish (Watanabe, 1993). Analysis of rotifers revealed that these HUFAs are not present in sufficient quantities in yeast fed rotifers, but are sufficient in those fed *Nannochloropsis* (Watanabe, et al. 1983a). Apparently this explains the nutritional inferiority of yeast-fed rotifers. A similar situation exists for rotifers cultured on freshwater *Chlorella*. Rotifers fed freshwater *Chlorella* grow well, but are nutritionally inadequate for many marine fish. The HUFA profile of freshwater *Chlorella* and marine *Nannochloropsis* are quite different and evidently *Nannochloropsis* better matches the HUFA requirements of most marine fish. *Nannochloropsis occulata* and *Isochrysis galbana* have a high content of essential fatty acid eicosapentaenoic acid (EPA 20:5ω-3) and docosahexaenoic acid (DHA 22:6ω-3) and are excellent microalgae for rotifer culture and enrichment.

Rotifer enrichment

A variety of feeding schedules have been examined for their ability to produce large quantities of rotifers of good nutritional quality. Rotifers mass cultured on yeast and then fed *Nannochloropsis* for the last two days before harvest would reach maximum levels, of HUFAs necessary for fish larvae. The high content of essential fatty acid, eicosapentaenoic acid (EPA 20:5ω-3) and docosahexaenoic acid (DHA 22:6ω-3) in specific microalgae (*e.g.* 30:5ω-3 in *Nannochloropsis occulata* and 22:6ω-3 in *Isochrysis galbana*) have made them excellent live foods for boosting the fatty acid content of rotifers. Rotifers fed these algae at a cell density of about 5 million cells/ml incorporate essential fatty acids in just a few hours and come to equilibrium with a DHA/EPA ratio above 2 (Dhert, 1996). Longer feedings does not increase HUFA levels. This secondary feeding approach to enriching the nutritional quality of rotifers just prior to feeding larval fish has become popular method used in many fish hatcheries. *Nannochloropsis* can be adapted to freshwater culture but may not be as nutritionally adequate as algae cultured in saltwater cultures.

A second approach to enriching rotifer nutritional quality is by feeding rotifers special strains of marine yeast that are high in omega-3 HUFAs. Unfortunately these yeast are slow growing and require expensive equipment for culture so they have not proven cost effective. A third approach is to add a HUFA emulsion (cuttlefish, polack, cod liver, menhaden, copepod, and halibut roe oil) to live baker's yeast cultures, allowing these cells to take up the fatty acids and lipid soluble vitamins before they are fed to rotifers (Watanabe, et al. 1983b). This is called the indirect method of enrichment and seems to work well. A fourth approach, called direct enrichment, is simplest. This method relies on the fact that rotifers can take up HUFA and lipid soluble vitamins directly from their medium. Rotifers are harvested, concentrated, and then soaked in a HUFA rich emulsion for 2-6 hr, thus becoming enriched with all the HUFAs required by larval fish (Watanabe, et al. 1983b). Enriched rotifers, however, lose their HUFAs quickly as they starve in larval rearing tanks. After about 6 hr in larval rearing tanks, rotifer HUFA content falls by about 2/3 and by 24 hr HUFA content falls to pre-enrichment levels (Jones, 1986). A procedure to avoid the poor nutritional quality of starved rotifers is to add enriched rotifers at frequent intervals or routinely add algae to the culture tank.

Enriching rotifers with lipid emulsions is not without problems. Rotifers immersed in these emulsions often clump together and form particles too large to be eaten by larval fish. Some enrichment emulsion is invariably carried over into the larval rearing tanks, potentially causing degradation of water quality. Commercial enrichments are expensive and if not properly handled, deteriorate quickly. Nonetheless, nutritional research has led to the steady improvement of these products so that they have become important components of rotifer feeds in many hatcheries. Many hatcheries, however, still prefer to enrich rotifers with microalgae because it is simple, reliable, does not degrade water quality nor does it encourage unwanted bacterial growth.

Rotifers can also be deficient in a variety of micronutrients and can be enriched for these (Hamre et al. 2008). Cod larvae cultured on bakers yeast fed rotifers grew slower and developed higher frequencies of deformities then copepod fed larvae. Similar culture results were obtained when rotifer HUFA and micronutrient levels were elevated to equal copepod levels. Algamac 2000 or *Chlorella* were sufficient for covering requirements of cod larvae for all the B-vitamins except thiamine. Rotifers cultured on Culture Selco had sufficient amounts of thiamine. Of the minerals, only calcium and magnesium were sufficient, while iron was on the borderline. Roti-Rich a yeast-based vitamin/mineral/micronutrient diet was not tested. This further illustrates that once fish larvae nutritional deficiencies are identified, they can be remedied by rotifer enrichment. Rotifers take up and encapsulate whatever they are fed, so their biochemical composition can be closely matched to the nutritional needs of fish larvae.

A recipe for a post-harvest enrichment emulsion was given by Watanabe, et al. (1983b). It consists of fish oil, methyl esters of $20:5\omega3$ and $22:6\omega3$ HUFAs, an emulsifier like lecithin, the fat soluble vitamins A and E, and seawater. A 1.7 ml volume of this emulsion is added per liter of concentrated rotifers (1000/ml) along with 0.2 g of baker's yeast per liter and the suspension is aerated for 12 hr at 25° C. Lipids and especially HUFAs must be fresh to be useful in larval nutrition. There is a strong tendency for these compounds to oxidize quickly in the presence of oxygen and become rancid. Additions of vitamin E, A and C often help slow deterioration. Proper storage of products containing high concentrations of lipids is extremely important.

To help emphasize the ability of rotifers to become nutrient laden biocapsules, consider the work by Merchie *et al* (1995). In this study they found that levels of vitamin C in rotifers directly reflects dietary intake both after culture and/or from enrichment. Rotifers cultured on a diet of baker's yeast and *Isochrysis* contain 148 mg vitamin C /g (dry weight) while algae fed (*Chlorella* & *Isochrysis*) rotifers contained 2289 mg/g after three days of culture. Enrichment of the algae fed rotifers for six hours did not enhance the vitamin C levels indicating maximum levels can be achieved on a diet of algae alone. In contrast, baker's yeast/*Isochrysis* fed rotifers increased to 1599 mg/g. Rotifers fed Culture Selco™ (vitamin C enriched) and *Isochrysis* had a level of 327 mg/g after three days and increased to 1559 mg/g after six hours enrichment. Combination diet of enriched Culture Selco™ and Protein Selco™

was comparable to baker's yeast at 136 mg/g and after six hours enrichment the level only increased to 941 mg/g. Storage of rotifers in seawater for 24 hours after harvest and enrichment did not significantly change vitamin C levels. However, these experiments demonstrated the value of microalgae as a sole diet or combined with enrichment.

It is more convenient to enrich the nutritional quality of rotifers with one of the commercial enrichment products combined with microalgae diet. These HUFA enriched products are formulated specifically for the nutritional needs of larval marine fish. Although expensive, these products are more consistent in quality and can be used to enrich the nutritional quality of *Artemia* and *Daphnia*. Some of these products are compared by Fernandez-Reiriz, et al. (1993).

Culture Equipment

Rotifers are easily cultured in containers with a variety of shapes ranging from two liter cola bottles to 20,000 liter tanks. Their container requirements are simple, but there are several points to remember when selecting a culture vessel. First, rotifer cultures must not be larger than your microalgae cultures can support. A larger tank is harder and more costly to feed then a smaller tank. Rotifers prefer a shallow (1 meter = 39" or less) vessel with a large surface area. Color of the container does not seem to be important however, for viewing during feeding, cleaning and harvesting a clear vessel is best. Second, the culture vessel should allow for good circulation of rotifer foods. Moderate to slow aeration provided in the back corners of a rectangular aquarium produces good circulation. Circular containers are ideal for circulation, but not as easy to harvest and maintain unless they are cone shaped with a valved drain on the bottom. Microalgae can also be cultured in a variety of vessels but in general clear, circular vessels are best (see microalgae section for details).

Figure 5.5 illustrates a simple, inexpensive, small scale, continuous culture system using two liter cola bottles for algae and 19 liter (5 gal) aquaria for rotifers. This system is a semi-continuous

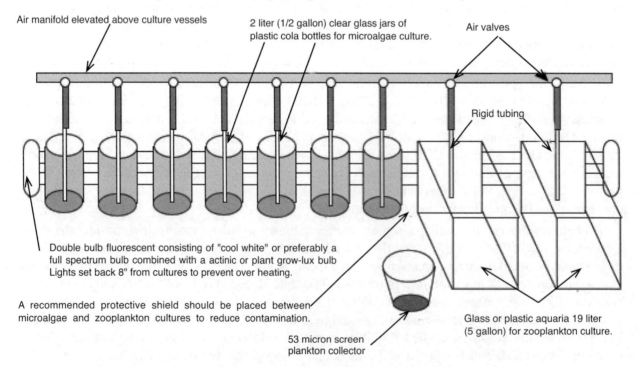

Figure 5.5. A small table top system for algae and rotifer cutlure.

culture where algae and rotifers are raised separately. This method contrasts with a batch culture where the algae is cultured for a short time and then a small quantity of rotifers is added directly to the algae culture and allowed to consume the food, harvested, then restarted.

Zooplankton harvesting equipment

To harvest all the rotifers, adults and juveniles, in a culture a 50-53μm screen (0.0020 inch) is best. Small hand held collectors are shown in Figure 5.6. Smaller mesh screens are fine but clog quickly when harvesting dense cultures. Tapping the sides of the collector realigns the material on the screen and speeds the collecting process. It is advisable that the screens be made from high quality polyester plankton cloth since is less incline to stretch and is resistant to deterioration in saltwater.

Plankton nets (Figure 5.6) are often used in collecting zooplankton in ponds, rivers and estuaries. These vary greatly in mouth size, length and mesh from the small example shown to very large cus-

tom made nets. In large ponds or tanks these nets are mounted into the cultures and a airlift pump or small, slow flow, magnetic drive centrifugal pump is used to pump water from the cultures into the collecting net. Where possible simple airlift pumps are more desirable to help prevent mechanical damage to the zooplankton.

Automatic collectors normally use gravity flow and flooded chambers to quickly and easily harvest large quantities of rotifers (Figure 5.7) (Misra and Phelps, 1992; Hoff, 1996). When harvesting large cultures it is important to make sure the collector has a flooded chamber to reduce mechanical damage to the zooplankton. As with the collector nets high speed electrical pumps are not good.

Figure 5.6 - Above - Hand held zooplankton collectors of various sizes and mesh. Left - Small zooplankton collecting net with detachable collecting containers (mfg. Florida Aqua Farms).

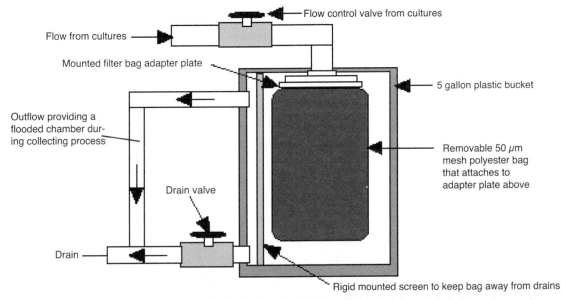

Figure 5.7. Design of gravity-flow zooplankton collector (Mfg. Florida Aqua Farms)

Procedure for Continuous Culture

The following procedure is scaled for the aquarist level of continuous rotifer production at 25° C. However, this design and procedure could be easily scaled up to larger aquaculture operations. The objective is to maintain a rotifer population in log growth phase and harvest at frequent intervals for a continuous supply of rotifers. In this culture regime, rotifers are fed microalgae supplemented with a yeast-based feed. The continuous culture procedure is outlined below (Table 5.3), followed by a more detailed description.

Procedure Day →	1	2	3	4	5	6	7	8	9	10
Start algae culture #1	1									
Start unmarked algae culture	1									
Start hatching rotifers			3							
Start algae culture #2				4						
Transfer rotifer culture					5					
Feed Roti-Rich 2x/day						6	7	8	9	10
Start algae culture #3						6				
Feed rotifer culture #1							7			
Re-bloom algae culture #1							7			
Feed algae culture #2									9	
Re-bloom algae culture #2									9	

Procedure Day →	11	12	13	14	15	16	17	18	19	20
Feed Roti-Rich 2x/day	11	12	13	14	15	16	17	18	19	20
Begin harvesting rotifers	11									
Feed algae culture #3	11									
Re-bloom algae culture #3	11									
Harvest rotifers			13							
Feed algae culture #1			13							
Re-bloom algae culture #1			13							
Harvest rotifers					15					
Feed algae culture #2					15					
Re-bloom algae culture #2					15					
Harvest rotifers							17			
Feed algae culture #3							17			
Re-bloom algae culture #3							17			
Harvest rotifers									19	etc

Table 5.3 — Procedure for continuous rotifer and algae culture.

Day One

A) Set up two, 2 liter clear, clean, plastic cola or glass bottles with sterile, premixed saltwater or fresh-water (see Microalgae Chapter for more details on preparing sterile culture water and culturing algae). Add 1.0 ml (1/4 teaspoon) per 2 liters of a f/2 nutrient media (Micro Algae Grow™) or an appropriate amount of another fertilizer to each bottle. Higher doses will not speed growth or produce higher yields and may impede growth. Some fertilizers use ammonia or urea as a nitrogen source and are not recommended for rotifer culture.

B) Label one microalgae culture Algae Culture #1 and leave other unmarked.

C) Add enough liquid algae culture or an entire Micro Algae Disk™ to produce a light green color. If using disks, cells should be removed by adding sterile culture water to the disk and rubbing off the cells with a "Q-tip." Better results are obtained by allowing the culture water to remain on the disk for 24 hours before removing the algal cells. If culturing in freshwater, blending the entire disk in a blender with sterile culture water may help. Refer to the microalgae section for complete details.

Day Two

A) Stir the algae cultures with a clean rod if the cells settle and turn up aeration. Stirring may be required twice a day for several days. As long as the cells are green the microalgae is normally alive.

Day Three

A) Place about 25 ml of unfertilized culture water in a small, clear container like a petri dish, add a vial of Resting Rotifers™ and cover. Rotifer cysts take 24-48 hours to hatch at 25°C with continuous, moderate light (about 4000 lux).

Day Four

A) Start another algae culture in a two liter container using a disk or an appropriate amount of liquid algae culture. Label this Algae Culture #2.

B) After 24 hours, check the rotifer cysts for hatching using a magnifying glass of about 8X+. Add about 4 ml or less of microalgae from the unmarked microalgae culture started on day one to obtain a light green color in the petri dish and cover This is done irregardless of whether or not hatched rotifers were seen.

Day Five

A) At 48 hours transfer the rotifer culture to a 500 ml (pint) container and feed more algae to obtain a light green color and a few drops of a invertebrate food (Roti-Rich™) Care must be taken not to over feed since this can impede growth and reproduction.

Day Six

A) Add rotifer culture to a 2 liter container, regardless if hatch was noted. Fill about half the container with microalgae from the unmarked Algae Culture, mix with new culture water to obtain a light green color. Caution, dense algae (a dark green color) elevates pH and slows rotifer reproduction. pH above 10 kills rotifers. Lightly aerate the rotifer culture container without use of a airstone. You are basically providing circulation not typical aeration.

B) Start another two liter algae culture using a disk or liquid inoculant. Label this Algae Culture #3.

Day Seven

A) If the 2 liter starter rotifer culture has cleared the algae, add more from the unmarked Algae Culture and 5 drops of Roti-Rich™.

B) Usually by 5-6 day the rotifer population reaches a size where they could easily consume 2 liters of moderately dense microalgae each day. At this point, supplementation should begin with small amounts of Roti-Rich™ or another comparable yeast-based food. Start at 1 ml, (20 drops or about 1/4 teaspoon) administered 1/2 in the morning and 1/2 in the evening to feed a dense, 4 liter rotifer culture each day. Observe how long it takes water to clear to determine if more Roti-Rich™ should be added. Caution: do not overfeed; it is very easy to degrade water quality with yeast to the point where the rotifer population crashes or grows slowly. A guide is to look at the water prior to the second feeding. If still cloudy skip the second feeding.

Day Eight

A) Transfer the two liter rotifer culture to a five gallon aquarium and add the remaining volume of the unmarked Algae Culture. Also add 3 liters of unfertilized culture water. Lightly aerate in one corner of the aquaria. If rotifer culture water is clear, add about 660 ml (3 cups) from Algae Culture #1 to rotifer culture.

B) Feed the rotifer culture Roti-Rich™ twice daily, or less if water is cloudy between feeding.

Day Nine

A) Add all but about 200 ml (1 cup) of the Algae Culture #1 to the rotifer culture. Refill Algae Culture #1 with sterile culture water and re-fertilize as on day one. Note: re-blooming an algae culture is usually successful up to 3 times before the culture vessel needs to be discarded and/or thoroughly cleaned and restarted. If you cannot thoroughly clean the algae from the inside walls of a culture vessel, it should be discarded.

B) Feed the rotifer culture Roti-Rich™ twice daily, or less if water is cloudy between feeding.

Day Ten

A) If rotifer density exceeds 50/ml you can begin harvesting. Note that high densities may be reached earlier, so harvesting could begin earlier. Rotifer density can be visually checked using an 8-12X magnifier.

B) To harvest rotifers, pour or siphon about 1/2 of the rotifer culture through a fine mesh plankton collector. Wash the concentrated rotifers off the screen directly into an aquarium or larval tank. Add only the rotifers, not their culture water. Pour the filtered water back into the rotifer culture vessel. When the five gallon aquarium is filled, remove the excess water before adding more microalgae. Harvest only every other day or reduce harvest to 1/3 for daily harvests.

C) Add all but 200 ml (1 cup) of Algae Culture #2 to the rotifer culture. Refill microalgae culture vessel as instructed on Day Nine (A).

D) Feed rotifer culture twice a day with a suitable liquid invertebrate food. (example -Roti-Rich™).

E) Algae growth rates depend on ambient conditions, health of the inoculant, water chemistry, and age of the culture. Algae cultures can be sustained by adding a dose of fertilizer every 7-8 days, but the nutritional quality of the algae deteriorates with age. Frequent re-blooming or use of new inoculants keeps the algae in its fastest growing stage and most nutritious condition. Refer to the microalgae section of this manual for more information.

Additional Points

A) This procedure uses three algae cultures which are re-bloomed every 6 days and one rotifer culture, half of which is harvested every other day. If rotifers are needed every day, this can be accomplished by maintaining three additional algae cultures and another rotifer culture or cutting down harvest to 1/3 each day. Extra algae cultures are recommended to protect against crashes and culture failures. Snell and Hoff (1989) showed by computer simulation of rotifer survival and reproduction, that daily harvest over a 25-30 day period is highest when only 20% to 30% of the population is removed daily. Lower daily yields (biomass) resulted when 10%, 40%, and 50% of the population is harvested.

B) Established rotifer cultures can be sustained for long periods on Roti-Rich™ of yeast alone, but reproductive rate is higher and nutritional quality better when microalgae is at least 10-20% of their diet.

C) Rotifer cultures can be easily maintained for months with a minimum of time and effort, providing the food is good, water quality is maintained, and the culture does not become contaminated with competing organisms.

D) Rotifer culture vessels should be cleaned every 3-6 weeks. Transfer between 10% and 20% of the water from the old culture into the new rotifer culture vessel. Filter the rotifers out of the remaining old water and discard the water. Place concentrated rotifers in a clean tank filled with new, aged water, and 5% to 10% new algae water. If the old culture water is foul smelling or contaminated with ciliates, concentrate the rotifers, wash them with freshwater while in the collector and place them in the new culture vessel with clean, aged water. Usually, initial survival immediately after transfer and the recovery rate is not as good when 100% of the water is changed.

Procedure for Small Scale Batch Culture

Batch culture follows a similar procedure as continuous culture, except that all the rotifers are harvested at once, the culture water discarded, the culture vessel sterilized and restocked with algae and rotifers. There are distinct advantages with the batch method. These include: 1) production is more predictable 2) there is less chance of contaminants overtaking the rotifer culture 3) it is easier to maintain rotifer populations growth in log phase. Batch culture is especially applicable in situations where a known supply of rotifers is needed daily. The following design, protocol, and results were obtained from work conducted at Florida Aqua Farms for tropical fish farmers (Hoff and Snell, 1988).

The system design (Figures 4.10 and 5.7) is scaled to the daily needs of a small tropical fish farm requiring about 30 million rotifers a day. The design is modular and can be easily scaled for higher or lower rotifer production. The following data were based on the culture of the freshwater rotifer *B. calyciflorus* over a 25 day period. Similar results were also obtained using the same system and procedure for the marine rotifer *B. plicatilis*.

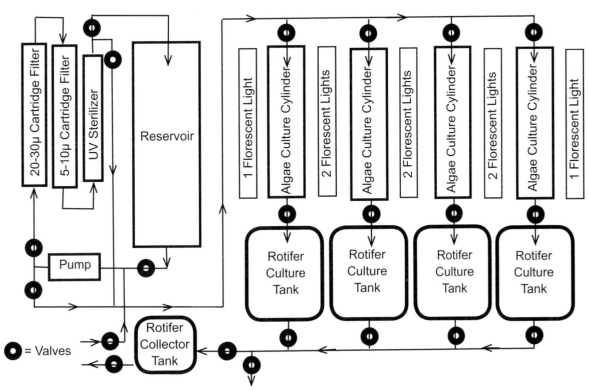

Figure 5.8. Commercial scale rotifer culture system flow diagram designed by Florida Aqua Farms. See Figure 4.10 for another drawing of this system.

System Design

This rotifer culture system consisted of nine, rectangular 83 liter (22 gallon) plastic laundry sinks (Figures 4.10 and 5.8). Eight were equipped with a removable stand pipe and plumbed into a central drain manifold which flowed into another modified sink for harvesting and concentrating rotifers. The collector consisted of a square 53 μm mesh plankton net on the bottom of the sink. The large central pipe served as a reservoir for retaining the concentrated rotifers and was equipped with a valve for draining. Culture water flowed through the net and out of the sink drain. As a result, rotifers were filtered, concentrated, and forced down into the reservoir. A more recent collector design is also shown in Figure 5.7 which utilizes a flooded container and removable collector bag (Florida Aqua Farms).

Above each sink was a 93 liter (25 gallon), 30 cm diameter x 122 cm high (12" dia x 4'), clear fiberglass algae culture cylinder. Each cylinder was equipped with a bottom valve situated so that algae could be directly drained into the rotifer culture sink below. Between each cylinder was a double fixture, 122 cm high (4') florescent light, mounted vertically on each side of the cylinder. Bulbs consisted of one "Cool White" and one "Actinic" or "Plant Grow" bulb per fixture. The lights were controlled by a timer, but photoperiod was normally set at 24 hr light.

Each rotifer culture sink and algae culture cylinder was supplied from an air manifold plumbed to an isolated air pump equipped with an intake filter. Moderate aeration without airstones was supplied to each algae cylinder and slow aeration was supplied in the center of each rotifer culture sink. Air lines to algae cylinders were equipped with check valves to prevent siphoning.

Culture water came from a well and was filtered through 35 μm and 5 μm cartridges, and then passed through a 40 watt UV filter and stored in two, 147 liter black, covered reservoirs. Water was then chlorinated with dry shock treatment chlorine powder, heavily aerated, and continually passed back through the cartridge filter and UV in a closed loop for 12 hours. After 12 hours the water was dechlorinated with sodium thiosulfate and tested for residual chlorine with chlorine test strips. If negative, the water was ready to use for microalgae culture. Use of a special treatment reservoir looks conceptually good but is harder to maintain and keep sterile. Filling the tanks directly from the filters and then chlorinating and dechlorinating in the algae culture tank is now preferred. Space requirements for this system was about 12'L x 6'W x 8'H. It was maintained in a temperature controlled room using a small 6,000 BTU air conditioner. Light was normally on for 24 hours. Air was supplied by a separate aerator maintained inside the room.

Algae Culture Procedure

1) Four liter microalgae starter cultures of Nannochloropsis were maintained in a separate temperature controlled culture room.

2) Eighty five liters of sterile water (see Treatment of Culture Water in the microalgae chapter) was pumped from the reservoir into an algae culture cylinder, fertilized, and inoculated with 3.5 liters of Nannochloropsis from a 7 day old starter culture (see Figure 4.16).

3) Starter culture was then refilled with sterile water, fertilized and allowed to re-bloom. Process was repeated 3-5 times before new cultures had to be restarted from another liquid subculture.

4) The algae cylinders were allowed to grow for 5 days with 24 hr light provided. On the fifth day, 75 liters of algae was drained into a rotifer culture sink, leaving approximately 12 liters of algae in the cylinder.

5) The algae cylinder then was refilled with sterile water, fertilized, and allowed to grow for another five days and the entire process was repeated. Best results were obtained when re-blooming was repeated only 3-4 times over 15-20 days. After the final re-blooming, the cylinder was removed, brush cleaned with a small amount of detergent water, rinsed, and acid washed to remove attached diatoms and calcium.

Algae production results

Algae mass cultures inoculated with 20 liters from an older microalgae culture yielded a starting density of 68,000 cells per ml and produced an average of 6 million cells per ml after 5 days at 25° C. Standard deviation was 1.8 million cell per ml. Mass cultures started from 3.5 liters new microalgae culture (log phase) averaged 5.3 million cells/ml after 5 days starting from an initial density of 15,000 cells per ml. Typically, new microalgae cultures are better for inoculation rather than older cultures.

Rotifer culture

1) Prior to inoculating rotifers, the pH of the algae culture was adjusted to 7.9 using small amounts of acetic acid. Note, this procedure has been modified since this study by adding 10-20% old culture water to the new culture. Acid is no longer required, but can be used if pH exceed 7.5.

2) After adjusting the pH, clean cultures of rotifers (those with few to no ciliates) were used to stock the mass culture at 50 rotifers per ml (approximately 3.8 million rotifers). The mass culture then was allowed to reproduce for 5 days.

3) During the five days, supplemental feeding of Roti-Rich™ and additional algae were provided as follows:

Day 0 = Stock algae, adjust pH, add rotifers.

Day 1 = 40 ml Roti-Rich (20 ml in AM, 20 ml in PM)

Day 2 = 60 ml (30 ml AM, 30 ml PM)

Day 3 = 60 ml (30 ml AM, 30 ml PM) + 5 to 10 liters algae in AM

Day 4 = 60 ml (30 ml AM, 30 ml PM) + 5 to 10 liters algae in AM

Day 5 = 30 ml in AM + 5 liters of algae, harvest rotifers, and restock culture tank.

4) Each day the tanks were stirred to mix settled algae cells back into suspension.

5) On day 5 all rotifers were harvested after feeding on microalgae for several hours. Culture tanks were then cleaned and restocked with algae, the pH adjusted, and inoculated with rotifers.

Examining rotifer populations

Rotifer populations should be monitored daily, especially if there are production problems. Observations of female and male density, number of females carrying one and multiple eggs, and ciliate contamination should be recorded. Although burdensome, it is essential for maintaining consistent production and for controlling problems before they get out of hand. With experience, these observations can be completed in 3 minutes per tank (see Monitoring Rotifer Populations).

Rotifer production results

A total of 25 experiments were conducted and the results of these experiments are summarized in Tables 5.4a and b. Contamination with large numbers of the ciliate *Euplotes* considerably reduced rotifer production. Good mass cultures averaged 475 rotifers per ml (39 million) in five days, while contaminated cultures averaged 125 rotifers per ml (10 million) (Figure 5.9). The presence of ciliates substantially reduced the number of females bearing single and multiple eggs (Figure 5.10), but not algal densities (Figure 5.10 and 5.11). Since microalgae levels were not suppressed by *Euplotes*, these ciliates reduced rotifer reproduction by some other means. Ciliates primarily feed on bacteria and apparently compete for yeast-based foods. Rotifers consume both bacteria and yeast as well which might supply them with essential nutrients. When these foods are limited due to ciliate contamination, rotifer reproduction is hampered. Another possibility is that ciliates could secrete inhibiting substances into the water that slows or directly effects rotifer reproduction.

Removal of ciliates

In rotifer mass cultures, two species of ciliates are common contaminants, *Uronema* and *Euplotes* (Figure 5.12). Filtering ciliates out of the rotifer inoculant can be accomplished using a 50 μm screen and flushing the rotifers with freshwater to dislodge attached ciliates. Rotifers do best when transferred into new culture water that has about 10% to 20% of the old culture water. Therefore, during transfer, old contaminated culture water (20%) should be filtered through a 5 μm filter or less to remove all ciliates. If a culture is contaminated, reducing the use of yeast-based foods helps reduce ciliate growth. *Artemia* feed on *Euplotes* and *Uronema* (Maeda and Hino, 1992) and can be polycultured with rotifers. Interactions of rotifers with ciliates and other zooplankton species commonly contaminating mass cultures have been described by Hagiwara, et al. (1995). Some protists, like the heliozoan *Oxnerella maritima*, make rotifer mass cultures unstable and can kill rotifers (Cheng et al. 1997). Higher rotifer densities (>100/ml) may not be achieved using polyculture, but where ciliate contamination is prevalent, screening may be a cost-effective control. Cheng et al. (2004) indicate that *Euplotes vannus* outcompetes *Brachionus rotundiformis* in *Tetraselmis tetrathele* cultures. However, if the diet is switched to *Nannochloropsis oculata*, *Euplotes* is suppressed and rotifers regain dominance in the cultures. Manipulating diet may therefore be an excellent way to control ciliates in rotifer mass cultures.

Observation	Day 0	Day 1	Day 2	Day 3	Day 4	Day 5
% Rotifers, single eggs	14	14	8	14	13	12
Standard deviation	7.3	15.4	4.5	5.6	7.4	5.8
% Rotifers, multiple eggs	1	31	15	5	2	2
Standard deviation	3.2	16.2	11.2	6.9	4.0	5.0
Rotifers per ml	50	90	231	366	431	475
Standard deviation	0	51	34	156	162	147
Algae cells/ml (milions)	5.8	1.8	0.9	0.6	0.3	0.07
Standard deviation	1.8	1.3	0.8	0.4	0.3	0.07

Table 5.4a - Production of rotifers not contaminated with ciliates. Egg bearing rotifers over a 5 day culture for B. calyciflorus. Rotifers are number per ml, algae are million cells per ml.

Table 5.4b - Production of rotifers contaminated with ciliates.

Observation	Day 0	Day 1	Day 2	Day 3	Day 4	Day 5
% Rotifers, single eggs	9	21	19	10	6	7
Standard deviation	5.8	9.9	12.4	5.0	4.0	5.7
% Rotifers, multiple eggs	0	14	7	2	0	0
Standard deviation	0	16.3	6.2	4.6	0	0
Rotifers per ml	50	47	100	145	138	125
Standard deviation	0	28	73	74	94	98
Algae cells/ml (milions)	4.7	1.9	0.7	0.5	0.2	0.1
Standard deviation	1.3	0.9	0.9	0.5	0.3	0.2

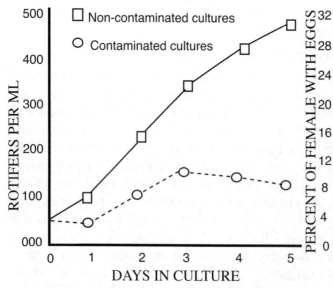

Figure 5.9 . Comparison of rotifer densities in contaminated versus non-contaminated cultures.

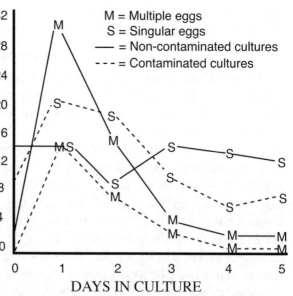

Figure 5.10 . Percent singular and multiple egg-bearing females in contaminated versus non-contaminated cultures.

Rotifer production costs

Based on this simple economical design, a hatchery could easily scale this unit to produce an average of 235 million rotifers per day using 500 liter (133 gal) culture vessels, 572 liters (152 gal) of microalgae at 6 million cells per ml and 1.1 liters (1.2 qt) of Roti-Rich™. Estimated daily cost for 235 million rotifers per day, including 4 hours labor per day at $8.00 per hour to maintain 7 culture tanks is <$57.00 or approximately $0.24 cents per million rotifers (based on a 5 day growout). Fatty acid enrichment would raise the cost to around $0.28 cents per million.

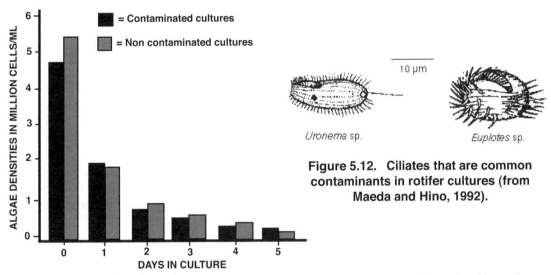

Figure 5.12. Ciliates that are common contaminants in rotifer cultures (from Maeda and Hino, 1992).

Figure 5.11. Algal densities in contaminated compared to non-contaminated rotifer cultures.

Procedure for Large Scale Batch Culture

In Hawaii the Oceanic Institute described a quick two day method utilizing 1,200L (317 gal) tanks that are half filled with algae at a density of 13-14 million cells/ml and inoculated at 100 rotifers/ml on day one. Salinity is 23 ppt, temperature at 30° C (86° F) and constant light. On day one activated baker's yeast is provided two times a day at 0.25g/million rotifers/feeding. On day two the tanks are completely filled with algae at the same concentration and 0.375g of yeast/million rotifers is added twice a day. Day three the rotifers are harvested and new tanks are inoculated.

The typical procedure for high intensity marine rotifer culture in Japan is a combination of continuous algae culture and batch rotifer culture (Fulks and Main, 1991; Fukusho and Hirayama, 1992). *Nannochloropsis* is the primary food and is grown in very large containers inoculated at algal densities of 5-10 million cells/ml. Under summer conditions of light and temperature (20-28° C or 68-83° F), cell density can double in 3 days. At lower temperatures (7-15° C or 45-59° F) cell density doubles in about 7 days. Fertilizer consisting of organic and inorganic nutrients is added every 5 days. The fertilizer consists of Marine G-1 (a viscous mix of various ground fish) at 20g/ton H_2O (1000 L or 264 gal), $(NH_4)_2SO_4$ 50g/ton, Urea 2.5g/ton, $Ca(H_2PO_4)$ 7.5g/ton, and chelated metal mix (CLEWAT 32) 1.5g/ton. The culture tanks are aerated vigorously. When the algae density reaches 20-30 million cells/ml, about half the volume is used to start a batch rotifer culture. The algae culture is then re-bloomed with fresh, filtered seawater and more nutrients added.

Algae cultures is adjusted to pH 8.0, temperature is brought to 30° C (86° F), and salinity is adjusted to 20 ppt (1.0145 density) before rotifer inoculation. Rotifers are then stocked at 200/ml. Supplemental food is added consisting of activated bread yeast dissolved in water containing 1 gram chelated metal mix (CLEWAT 32™)/10 L. The metal mix serves as a buffer and provides suspended trace metals to the yeast. The yeast solution is dripped into the culture tank at a constant rate. After 48 hours the rotifer population reaches a density of 700-1000/ml.

Within 24 hr after rotifer inoculation, *Nannochloropsis* density drops from 15 million to about 200 thousand cells/ml. Within 48 hr algae density falls to essentially 0. Supplementary feed is 120 g yeast/ton H_2O (0.12 g/L) for the first 24 hours increasing to 500 g (0.50 g/L) on following days.

Chemostat Mass Cultures

An example of the chemostat or continuous culture approach to microalgae and rotifer mass culture is diagramed in Figure 5.13. The primary advantage of this method is tighter controls on contamination and water quality, control of daily production, lower volumes of water and microalgae, and labor savings. James and Abu-Rezeq (1989a and b) described a chemostat system where the standard culture volume in their system was 1 m³ (266 gal). A diet of *Nannochloropsis* was provided at 20 million cells per ml and baker's yeast at 0.3-0.4 g per million rotifers per day. The microalgae were cultured in a separate chemostat maintained at about 50 million cells per ml (James, et al. 1988) and then diluted to the appropriate density in a mixing reactor before being fed into the rotifer tank. A harvest rate of 500 liters (133 gal) per day was used for L strain rotifers from a 1 m³ (1000 liters or 264 gal) culture chemostat each day. Average production from this system is 187 million rotifers per day (376 per ml), which has been operated continually for several months without problems.

Production from 1 m³ (264 gal) chemostats is currently sufficient to meet the rotifer needs of most small to medium size hatcheries. For example, to produce 1 billion rotifers per day requires 6-7 one m³ chemostats with the yields cited above. This means that 6-7 m³ (1584-1848 gal) of water can be managed to produce 1 billion rotifers per day, which is about 100-fold less than the volume of many batch culture methods currently used in Japan (Fukusho, 1989a). Intensively managing much smaller volumes of water results in substantial labor and cost savings for the chemostat method. In addition, the nutritional quality of the rotifers can be more tightly controlled. James and Abu-Rezeq (1989a) reported that the ω3 fatty acid composition of chemostat raised rotifers provided adequate quantities of essential fatty acids for marine fish larvae without further enrichment.

Very large marine fish hatcheries require the production of about 20 billion rotifers per day (Fulks and Main, 1991). The chemostat method could supply this biomass of rotifers, but it would take 100-140 one m³ tanks. To meet the rotifer requirements of very large hatcheries, 10 m³ (2640 gal) chemostats would be preferable, so that 12-14 could yield 20 billion rotifers per day. 10 m³ chemostats represent a 10-fold upscaling of existing systems. However, the basic principles of chemostats are well developed, so a modest amount of research should provide the information necessary to accomplish this task. James and Abu-Rezeq (1989b) reported no differences in production when upscaling 100 L chemostats to 1 m³ size. This approach, therefore, offers a good opportunity for reliably supplying high quality rotifer biomass for larval culture.

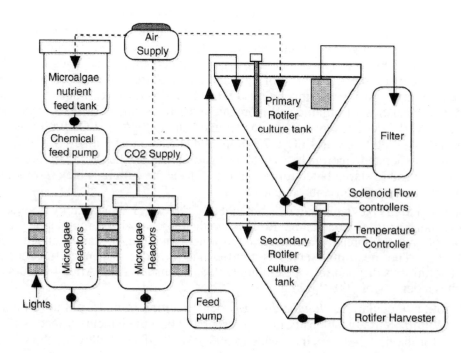

Figure 5.13. Chemostat culture unit.

High Density Mass Culture

In the 1980s, typical Japanese marine finfish hatcheries produced 1-20 billion rotifers per day at densities of 100-300/ml in 100-1000 m³ tanks (Yoshimura, et al. 1997). Two to three times more tank volume was required to grow the microalgae to feed these rotifers. As a consequence, 70-90% of total hatchery tank volume was used to produce live food, leaving only 10-30% for rearing fish larvae. Live food production was therefore placing severe limits on the number of fish larvae that could be reared.

By the early 1990s, it was discovered that condensed freshwater *Chlorella* cells remain alive and nutritious for several weeks if they are refrigerated (Hagiwara et al. 2001). Condensed *Chlorella* then was commercially introduced in Japan as a substitute rotifer food for *Nannochloropsis* (Hirayama, et al. 1989). When hatcheries were relieved of the necessity of growing microalgae, the amount of tank space devoted to live food production could be dramatically decreased. Condensed *Chlorella* allowed rotifers to be cultured at very high densities of 10-25 thousand per ml, cutting live food production labor by 2/3, costs by 1/3, and space requirements by 75% (Yoshimura, et al. 1997). A summary of the methods for this significant advance in rotifer mass culture is presented below.

The high density culture system is composed of 1 m³ tanks that are operated as semi-continuous cultures harvested at two day intervals (Figure 5.14) (Yoshimura, et al. 1997). *Brachionus rotundiformis* are inoculated at 10,000 individuals/ml and grow to 25,000/ml in two days at 25° C, salinity 25-30 ppt. These extraordinarily high densities are possible because the condensed *Chlorella* supplies sufficient

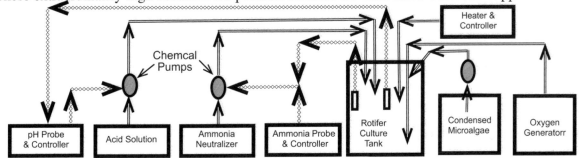

Figure 5.14 . Basic system design high density rotifer culture.

food with minimal degradation of water quality. Little degradation occurs because the *Chlorella* cells are alive and retain potentially toxic chemicals like free ammonia within their cells. Dead cells leak compounds that eventually reach concentrations that are toxic to rotifers. Consequently, it may not possible to achieve these very high rotifer densities using dry yeast or algal feeds.

Condensed *Chlorella* at a concentration of about 15 billion cells/ml is supplied continuously using a peristaltic pump. Consumption rate is approximately 0.5 L of condensed *Chlorella* per 10^8 rotifers per day. When rotifer densities reach 10,000/ml, deficiencies of vitamin B_{12} can develop (Maruyama and Hirayama, 1993). To avoid B_{12} limited rotifer growth, B_{12} should be provided at 200 μg per 100 g dry weight of *Chlorella*. Alternatively, vitamin B_{12} can be provided in the rotifer culture medium at 150 ng/L (1 ng = 10^{-9} g).

At very high rotifer densities, other factors like oxygen can become limiting. Providing rotifer cultures with oxygen increases production 3-4 times (Yoshimura, et al. 1994). When O_2 gas is used, it is typically supplied at 1-5 L per minute. Free ammonia secreted by rotifers as a waste product of nitrogen metabolism can reach toxic concentrations. A 50% reduction in rotifer population growth rate and fecundity was observed at free ammonia concentrations of 8-13 mg/L (Yu and Hirayama, 1986). Ammonia is considerably more toxic in its free (unionized) form (NH_3) than ammonium (NH_4^+). At higher pH, more ammonia is unionized and therefore toxic than at lower pH. Consequently, it is beneficial to maintain high density rotifer cultures at pH 7 by controlled additions of hydrochloric acid.

When rotifer density exceeds 1000/ml, feces rapidly accumulate in the culture tank. This heavy load of particulate organic matter can clog the plankton nets used to harvest the rotifers. A solution has been to suspend nylon filtration mats in the rotifer culture tanks to trap the organic particles. Daily removal and rinsing of the mats also helps to control protozoan parasites like *Vorticella* and helps stabilize the rotifer cultures (Yoshimura, et al. 1997). Nylon air conditioning filters work well as filtration mats, but they should be soaked in seawater for a few days before use and checked for toxicity. An improved design for the filtration mat is presented by Yoshimura, et al. (1997).

Aeration of the water in high density cultures can cause foaming. Foam sometimes overflows from the tank causing losses of up to 10% of tank volume and 50% of rotifer biomass (Yoshimura, et al. 1997). This loss can be minimized by using foam removing agents. A food grade, silica-based defoamer used at 10 ml/m^3/day will probably work. Several types of defoamers are available, but each needs to be checked for toxicity before use in mass cultures. Because of the potential toxicity of defoaming agents, they should only be used if foaming becomes a serious problem.

Fecundity is directly related to density levels. At a density of 150/ml egg bearing rotifers are about 30% of the population, however, at a density of 2000/ml the egg ratio drops to 10% and at 5000/ml it drops to 5%. Maintaining cultures with this low egg ratio is risky and demands rigid controlled conditions (Dhert, 1996).

Techniques for high density rotifer culture have been extended to the freshwater *Brachionus calyciflorus* (Park, et al. 2001). Using five liter batch cultures at 28°C with O$_2$ supplied and a diet of condensed *Chlorella*, densities of 17 to 19 thousand rotifers per ml were obtained. When pH was controlled to 7 and temperature increased to 32°C, a maximum rotifer density of 33.5 thousand rotifers per ml was achieved.

Pond Culture In China

In north China, along the Bohai Bay, great natural, brackish water populations of *B. plicatilis* appear in April around certain villages near aquatic or poultry processing plants. For years they were harvested by fishermen and sold to local hatcheries for larval culture of the marine shrimp, *Penaeus chinensis*, and the freshwater crab, *Eriocheir sinensis* (Fengqi, 1996). Demand exceeded supply and methods were devised to culture rotifers in earthen lined ponds. It was noted that rotifers can survive and reproduce very well in highly eutrophic conditions. The pond method described is based on this since they found that natural predators of rotifers, namely copepods, small shrimp and fish were eliminated by the eutrophic conditions. In natural waters with low nitrogen and phosphate it is almost impossible to eliminate copepods in earthen ponds. Using this knowledge artificially create highly eutrophic waters by applying fermented bean cake and chicken manure.

Fairly large rectangular ponds of 1,000 sq. m. surface area and 1 m. deep are used. Normally 5-10 ponds are required to meet their need for rotifers. Desired salinity is 10-20 ppt. so ponds are constructed near freshwater sources. In early spring as soon as the ice melts, sea water is pumped directly into the earthen ponds. Disinfecting the ponds is not required. About 500 kg of fermented bean cake and chicken manure in a 1:4 ratio is applied to each pond (Fengqi 1996). The dominant marine algae and bacteria species soon bloom. At the end of March at 15° C natural populations of rotifers appear and are stocked in the ponds at an initial level of 1-5 rotifers per liter. As the pond is used over and over wild rotifers are not needed because of the large number of resting eggs on the bottom. Only about 25% of the ponds are used for rotifer culture and 75% for algal culture.

Resting eggs on the bottom of the ponds will not hatch where dissolved oxygen is low and light is not sufficient (Hagiwara et al.1989). When resting eggs have been establish in the ponds they drain the pond in early spring exposing the bottom to the bright sun in order to oxidize the sediments. About 2-3 days later brackish water (10-20 ppt) is added to a depth of 20-30 cm, so that the resting eggs on the pond bottom receive sufficient sunlight to hatch. Numerous cysts also float to the surface (Hagiwara et al.1985). Cysts with about 1 m (39") water depth do not hatch so water depth is critical. After about 72 hours most of the cysts have hatched and algae water from another pond is added to the rotifer culture pond(s). Fresh or brackish water with additional fermented bean cake is added to the algae pond to sustain the culture. After 10 days in early April the rotifers are ready for harvest.

Typical yield is 100,000 rotifers/L (100/ml) at 15° C. A complete pond yields about 10 billion per day per pond. By adding algae continuously and exchanging a portion of the pond water daily, production can be extended for a month or longer (Fengqi, 1996).

Mother Cultures

Many hatcheries need mass amounts of rotifers seasonally or intermittently. This is especially true when working with pelagic spawning fish that generally have high fecundity releasing thousands of eggs per spawn. This up and down pattern of need and rest creates scaling problems in live food culture.

As noted in large scale culture techniques you would need 50-500/ml rotifers or more to even start a fast growing mass culture. Unlike Artemia, obtaining a can of rotifer cysts and hatch what you need each day, is not feasible. However, maintaining year round "stock cultures" provides a ready source from which you could scale up in a relatively short time. Live rotifer cultures are available from other sources but you also take the risk of inheriting their problems and contaminants.

Stock cultures require minimal labor and time. For over 10 years we have maintained stock cultures in a small air conditioned room in a hatchery. Using this method and feeding schedule produces a slow growing rotifer culture with only 10 to 15% of the population having singular eggs. The maintenance procedure is designed to meet only the basic needs of rotifers.

Maintaining Stock Cultures

1. Tank selection would be based on the volume of rotifers you would need to start a large scale tank. Generally mother cultures can be easily maintained at 100 to 200/ml.

2. Transfer 10% of the old rotifer culture water to the tank.

3. Harvest rotifers and stock the tank with 10 to 50 per ml.

4. Fill with new water and enough microalgae to create a medium green color.

5. From that point on, feed two times a day just enough yeast or yeast based foods like Roti-Rich to provide a light cloud in the water. If the cloud is present at the second feeding do not feed. Overfeeding is obvious based on a prolonged cloudy tank, hair-like growths around the air supply or along the tank walls.

6. Increase the amount of food per feeding based on how fast it is cleared by the rotifers.

7. Once a month transfer 10% of the old water to a new tank, harvest the rotifers and restart.

8. If you maintain microalgae sub-cultures the excess algae can be fed periodically.

9. At times the population will be too dense and thinning maybe required to keep it from crashing.

Monitoring and Assessing the Health of Rotifer Cultures

Since rotifers are small, determining rotifer density can be difficult without magnification of at least 8X. Magnifications of 10-15X make rotifer observation easier. When a culture is checked it should be well stirred, but not to the point where bottom detritus is pulled into the water column. Too much detritus makes it difficult to make rotifer counts and behavioral observations. If working with high densities (>400/ml), samples should be diluted for easier counting. Strong lighting from below (substage illumination) provides the best conditions for observing. Rotifers are translucent white and are most visible against a dark background. A dark background can be obtained with most microscopes by manipulating the angle and intensity of the illumination.

Estimating Rotifer Density

1) Remove a 100-200 ml sample from each, well-mixed, rotifer tank.

2) With a one ml pipet, stir the sample to provide good mixing and remove a 1 ml sample.

3) Pipet a 0.1-0.5 ml sample into several wells of a multi-well plate. A clear, glass Boerner slide with ten wells is ideal for counting. Black porcelain well slides with overhead lighting also work. When filling chambers hold the pipet vertical to the bottom of each well and gently move your thumb to release the water. This prevents splashing or adding too much water too quickly.

4) Check the live sample under 15-30X magnification for contaminants, especially the ciliate Euplotes, is flat, oblong (60 µm long and 30 µm wide) and swims faster than rotifers in a twisting motion.

5) Examine the live sample for swimming speed of the female rotifers and for the presence of males. Males are small, clear, fast swimming, and have a black dot towards their posterior. Sustained populations of males and ciliates are indicators of impending production problems (see Additional Points).

6) After the live samples have been examined for activity and contaminants place a drop in each well of either 10% formalin, denatured alcohol, or vinegar to immobilize or kill the rotifers. This makes density counting and assessment of egg production easier.

7) Count the total number of rotifers in the one ml sample and multiply this times the number of ml in the culture vessel to estimate the total population.

8) Record the number of females with single and multiple eggs. High percentages of females bearing singular or no eggs are indicative of a slowly growing culture. The causes could be lack of food, poor water quality, sexual reproduction, and/or the presence of contaminants. High percentages of multiple egg bearing females indicate much faster population growth. Tables 5.4a and b show the average egg bearing females per day till harvest for the freshwater rotifer B. calyciflorus (Hoff and Snell, 1988).

9) A quick density estimate can be obtained by holding a 1ml pipet containing a rotifer sample up to a light and counting rotifers without magnification. This takes practice, but acceptable estimates of rotifer density can be obtained this way if rotifer densities are 5-30 per ml, however, ciliate contaminants are not usually visible.

Assessing physiological condition of cultures

Two techniques for assessing the physiological condition of rotifer cultures were described by Snell, et al. (1987). The first is swimming activity, which is measured by observing rotifer swimming over a grid with 1 mm squares. Although these authors described a quantitative test, swimming activity is a useful indicator of stress even when unquantified. With a little experience it is possible to detect slower swimming simply by observing females in a culture. If they appear sluggish as compared to healthy, rapidly growing cultures, physiological stress is likely.

Swimming activity is one of the most sensitive indicators known for rotifers. Slower swimming is the initial warning sign of physiological stress in a rotifer mass culture. This has been illustrated accessing swimming activity and its response to unionized ammonia. Swimming activity at 25°C declines linearly as unionized ammonia concentration increases, so that at 2.3 mg/L (ppm), swimming activity is 50% of the control. This concentration is nearly 10 times lower than the LC_{50}, the concentration of unionized ammonia required to kill 50% of a population in 24 hours. Other experiments have shown that the suppression of swimming activity reaches its maximum after only 15 minutes of exposure to ammonia (Snell, *et al.* 1987). Korstad et al. (1995) have applied this technique to assess the status of rotifer mass cultures in a marine finfish hatchery.

A second technique for assessing the status of rotifer cultures is egg ratio. Egg ratio is the number of eggs carried by females divided by the number of females. It has the advantage of requiring only a simple population count to determine, but the disadvantage of having an 18-24 hour time lag between the onset of physiological stress and a depression of the egg ratio. Egg ratios are useful for predicting future reproductive output of rotifer populations as long as a few guidelines are followed. Egg ratios typical of exponentially growing (log phase) *B. plicatilis* populations at 25° C ranged from 0.5-1.2 eggs per female (Snell, et al. 1987). Rotifer populations reproducing at a replacement level (stationary phase, an equilibrium with little increase or decrease) had egg ratios between 0.13 - 0.5. Once egg ratio fell below 0.13, however, populations declined. The sensitivity of egg ratio to unionized ammonia was lower than swimming activity, with 7.3 mg of unionized ammonia required per liter for a 50% reduction in egg ratio. Monitoring of swimming activity and egg ratios should make it easier for aquaculturists to keep rotifer cultures stable and productive.

Mass culture instability

An important unresolved problem in rotifer production is mass culture instability. The causes of this instability have been examined by Hirayama (1987) where he summarizes the most likely factors. Poor nutritional quality of rotifer food is often a problem, especially when yeast is used exclusively. *B. plicatilis* has a vitamin B_{12} requirement and rotifer production suffers when deficiencies arise. Poor water quality caused by decaying, uneaten food and accumulation of metabolic wastes can be limiting, particularly unionized ammonia, which is toxic to many aquatic animals. Recent studies on nitrogen flow in rotifer mass cultures are providing insight into the mechanisms of water quality changes (Aoki, et al. 1995; Aoki and Hino, 1996). Sexual reproduction and cyst formation reduce rotifer population growth because dormant embryos are produced that temporarily drop out of the population (Snell, 1987). Diatom blooms can interfere with rotifer feeding and ciliates, if they become dominant (Figure 5.8), can suppress rotifer populations (Reguera, 1984; Hoff and Snell,1988). It also has been suggested that viruses might play a role in rotifer mass culture crashes (Comps, et al. 1991). They described a Birna-like virus that was associated with dying rotifers in a poorly growing culture. Viral lesions were documented using electron microscopy and secondary bacterial infections were also noted. This type of virus is important in fish pathology in salmonid fishes as well as seabass, so there is concern that fish may become infected by eating infected rotifers.

When cultures become unstable, they can crash, reducing larval food production and causing larval starvation. Time required for reestablishing large rotifer mass cultures is significant, so the crash of a mass culture can seriously impair the ability to meet larval feeding demand for several weeks. If indicators of an impending crash could be developed, they would provide an early warning of trouble. Measures then could be taken to avert the crash and return the culture to stability. One hope for improved management of rotifer production systems is the development of good models that capture the principle dynamics of mass cultures. Alver et al. (2006) have developed an excellent example with their individual based model built from energy budgets. Using the model, it is possible to accurately simulate steady-state egg ratios, daily yield, maximum net growth rate, and optimal schedule of enrichment.

Storage of Rotifer Biomass

Short-term Live Storage

It is not always possible to match rotifer production with the food demands of larval fish. Sometimes there is an abundance of rotifers and few larvae and sometimes larvae go hungry because of inadequate rotifer production. The peaks and valleys of rotifer availability can be smoothed by storing the rotifer biomass for short periods in a refrigerator. Short-term storage can be accomplished following a protocol described by Lubzens, et al. (1990). Rotifers are concentrated to 1000 per ml and refrigerated at 4° C in the dark. Lubzens et al. placed 100 ml aliquots of rotifer concentrate in 250 ml flasks in 10 ppt seawater. Every other day rotifers were fed 0.15 g baker's yeast per 100,000 rotifers and the medium was changed every 4 days. After 31 days of this treatment, 65% of the rotifers were still alive. The nutritional quality of refrigerated rotifers is not likely to be optimal, so enrichment with algae or fatty acid emulsions prior to utilization is advised. Research on storage of rotifer biomass at supercooled temperatures (-0.5°C) may eliminate the need for handling rotifers during storage.

Long-term Freezing

Long-term storage can be accomplished by freezing concentrated rotifers in low salinity water (7 ppt, 1.0046 density). Prior to freezing, the rotifers should be fully packed with algae and other nutritional supplements like fatty acids and vitamins. An ice cube tray is ideal for this purpose. Freezing kills rotifers, so upon thawing they will settle out of the water column quickly. Consequently, good circulation must be provided to maintain the dead rotifers up in the water column where they can be seen and captured by larval fish. Frozen rotifers are only about half as nutritious as live animals, and are not as readily accepted. This technique was used at Instant Ocean Hatcheries for clownfish larvae with moderate success (Hoff, 1996). With special techniques, rotifer embryos can be cryopreserved in liquid nitrogen indefinitely (Toledo and Kurokura, 1990).

Disinfection of Rotifers

A large and diverse population of bacteria develops in all rotifer mass cultures. Although many of these bacteria are beneficial, some may be pathogenic. At times, therefore, it is desirable to disinfect rotifers to prevent transfer of pathogens into larval rearing tanks. Adult rotifers can be disinfected by dipping them into an 8.5 mg/L solution of Chloramine-T at 8° C for 3 minutes (Whyte, et al. 1994). Rotifers are then rinsed and transferred to sterile seawater. The effect of four other antibiotics on the growth of *B. plicatilis* has been reported by Yamauchi (1993). All antibiotics should be used very sparingly. Frequent exposure of bacteria to antibiotics selects for resistant strains that can cause more harm than good.

Rotifer cysts also can be disinfected to produce bacteria-free populations (Hagiwara, et al. 1994; Douillet, 1998). Hagiwara, et al. sterilized cysts by rinsing with sterile seawater several times, followed by soaking in 0.5 ppm sodium hypochlorite for 60 minutes, then 0.25 ppm for 30 minutes. Cysts were then rinsed with sterile seawater. Doulliet described a similar method using a 0.5% sodium hypochlorite solution (NaOCl) prepared by diluting commercial bleach (9.5 ml of 5.25% bleach diluted with 90.5 ml water). Rotifer cysts were then exposed for 3 minutes at 25° C and 15 ppt. The cysts were rinsed with sterile seawater and hatched as usual under sterile conditions. The resulting rotifers were bacteria-free and can be used to inoculate sterile cultures. Rotifers cultured on algae in the absence of bacteria do not grow as well (Hino, 1993, see section on Bacteria in Rotifer Cultures).

Trouble Shooting

1) Rotifer populations sometimes are slow to become established in new or freshly cleaned culture vessels. In well-established tanks, attached algae and bacteria act as stabilizing forces through complex interactions that are poorly understood. These effects may be due to the addition of vitamins or growth factors to the medium that are required by rotifers. Additionally, algae and bacteria recycle metabolic wastes, acting as biofilters in the culture tank. It is therefore not advisable to over clean culture vessels between batches. In addition, we have found that inoculating rotifers into a new culture with 10-20% of the old culture water helps reduce shock, initial mortalities and inoculates the beneficial microorganisms associated with rotifer cultures.

2) At times, new rotifer cultures do not reproduce as quickly as expected. Inoculating rotifer cysts or live rotifers into high algal densities is often the cause of this problem. High algal densities raise the pH above optimal levels, thus killing or slowing rotifer reproduction. Better initial survival is achieved using microalgae with cell densities less than 1 million cells per ml. (light to medium color in culture tank) combined with a 10-20% addition of old rotifer culture water and new water. Lowering the pH of an algae culture to 7.5-8.0 with acetic acid before inoculating rotifers may help.

3) Heavy aeration causes excessive turbulence and can strip rotifers from a culture. Rotifers are easily trapped in the foam that builds up on the sides of the culture vessels. The minimum amount of aeration necessary to keep food particles in suspension is all that is required. Airstones should not be used in rotifer cultures.

4) Different size rotifers can appear seasonally in a process called clonal replacement. Clonal replacement occurs as one strain replaces another as conditions in ponds or culture tanks change from spring to fall. Different rotifer strains can vary in size and shape and these differences appear to be genetically based (Snell and Carrillo, 1984). Another phenomenon is called cyclomorphosis, where individuals in a single strain change morphologically through the seasons. In general, size increases in the winter months.

5) Other possible reasons for culture problems are:

   a) Insufficient inoculant size for the culture vessel - success of a culture depends on a certain number of rotifers multiplying at a particular rate when supplied with a particular food density. It is important to start your culture according to the directions.

   b) Insufficient food levels - too little yeast or Roti-Rich™, low algae densities, or contaminated algae cultures can contribute to failure.

   c) Ciliate contamination - ciliates can be reduced by filtration with a 50 μm screen and rinsing concentrated rotifers with freshwater. This allows ciliates to pass through, but retains the rotifers. The ciliate *Euplotes* can be controlled in algae cultures with formalin. Formalin is added to the algae culture until it reaches a concentration of 20 ppm at least one day prior to stocking rotifers. The growth of *Nannochloropsis* usually is not reduced at this level of formalin.

   d) Since rotifers tolerate high and low temperature and salinity levels these attributes can be used to eliminate many contaminates, but should not be maintained for extended periods since rotifer reproduction will cease or be greatly curtailed.

   e) Too much food - high algae levels or over feeding of yeast or Roti-Rich™ can cause water quality problems. We have found daily feed additions rather than trying to maintain a constant food level is easier to control and produces similar densities of rotifers in smaller cultures.

   f) Rotifer inoculant is in a nonreproductive state - it is important to use well fed, active rotifers or rotifers from newly hatched cysts each time you inoculate a new culture.

   g) Very low aeration results in excessive fouling while high aeration causes foam and stripping.

   h) Use of an inappropriate algae species as food - poor population growth.

   i) Un-ionized ammonia levels should be below 3 mg/L.

j) pH should be kept below 8.0, preferably 7.0 - 7.5.

k) Oxygen levels below 1.2 mg/L can be lethal. Reduce suspended organic particles by placing vertically mounted filter-pads or brushes into the culture tanks and clean daily. Increase the surface area verses depth ratio of the tanks.

l) Excess foam from over feeding can be controlled with a silica defoamer. Switch from active yeast to inactive yeast.

m) Copepod contamination. Unlike rotifer cultures contaminated with ciliates and protozoan, copepods are predators of rotifers. Increasing the ammonia levels or creating eutrophication conditions should eliminate copepods and others predators such a larval fish, shrimp and crab larvae. Copepods are also susceptible to elevated phosphate and nitrate.

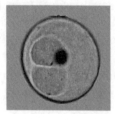

2 cell stage - 1:20 hrs

Prehatch - 75:35 hrs

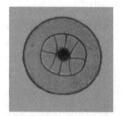

8 cell stage - 2:20 hrs

Late neurula - 61:20 hrs

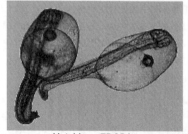

Hatching - 75:35 hrs

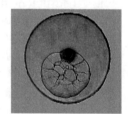

32 cell stage - 4:15 hrs

Mid neurula - 37:05 Hrs

Photographs by: Frank Hoff &
Joe Mountain,1972.

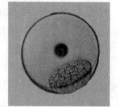

64 cell stage - 5:05 hrs

Development stages of the Black Sea
Bass (Centropristes melanus) at 19°C
(66°F).

Late blastule - 26:50 hrs

From: Conditioning, Spawning and Rearing of
Fish with Emphasis on Marine Clownfish (Hoff
1996).

Chapter 6 - CILIATE CULTURE

Introduction

Role in aquatic environments

Little is known of the role ciliates play in aquatic food webs. Yet, based on their abundance, it is highly likely that they are an important food for the early life stages of many fish and invertebrates (Stoecker & Cappuzo, 1990). Ciliates are members of the phylum Cilophora and are characterized as unicellular microorganisms that have hair-like cilia covering of covering their surface.

One of the most popular ciliate genus is *Paramecia*. Most of us remember our biology class where the teacher cultured some paramecium for us to view under a scope. Paramecium are a cosmopolitan organism and are found in suitable habitats all around the world. Global distribution of *Paramecia* species is believed to be the result of the break-up of the super-continent Pangaea over 200 million years ago. This continent was home to ancestral paramecium that have subsequently been separated by continental drift. Oldest reported fossil paramecium were discovered in amber dating back to the Cretaceous period, over 65 million years ago.

Most ciliates including paramecium feed on bacteria (bacteriovorous). Bacteria feed on decaying organic matter. Ciliates serve as an important link in detritus-based food webs in aquatic ecosystems which are in turn preyed upon by larger organisms. Ribblett and Coats (2000) reported on ciliates in freshwater streams. Ciliates are very abundant on decomposing sycamore leaves, with densities exceeding 25,000 ciliates per leaf and a diversity of over 50 species that inhabit decomposing leaves. Of these most consume bacteria, but a few are specialized predators that feed on other protozoans while others are omnivorous. Within streams the role of protozoa in transfer of energy from bacteria and fungi associated with decomposing leaves to higher trophic levels remains unknown.

Tintinnid ciliates are known to be consumed by wild fish larvae. Biological information and mass rearing techniques are discussed by Taniguchi (1978). Yet, more specific information on utilizing ciliates as food in fish and invertebrate culture is not widely available. Popular articles in aquarist magazines have reported the use of ciliates as food for rearing damsel fish, but details are lacking on success rate and methods employed. Experiments conducted on clownfish (*Amphiprion*) at Instant Ocean Hatcheries indicated that ciliates were consumed by newly hatched clownfish, but the nutritional value was not determined (Hoff, 1996). However, Howell (1972), reported that the ciliate *Euplotes vannus* was not accepted by lemon sole larvae. Gealy (2001) stated that paramecium are an excellent primary food source for newly hatched Australian rainbowfish (*Melanotaenia sp.*) larvae.

Infusoria (single-celled organisms consisting mainly of ciliate protozoans) is a common word in popular hobby literature to designate a paramecium culture. However, this is really an encompassing collective word that may include mixtures of phytoplankton and zooplankton microorganisms like ciliates, microalgae, bacteria, protozoans desmids, rotifers etc.. An early method for obtaining a infusoria culture consisted of boiling hay, lettuce, spinach or other vegetable matter and allowing the resultant infusion to stand in the air for a while in the hope that stray infusoria will alight therein. In most cases it did not result into a successful culture.

Cryptocaryon irritans, a common parasitic ciliate that is well recognized as a serious pathogen in warm water mariculture. Steidinger et al. 1996, Burkholder et al. 2001a studied apostomes for many years which are ciliates that are symbionts commonly associated with decapod crustaceans. The predatory ciliate *Didinium* consumes other ciliate protozoa and is considered a "lion" in the world of zooplankton. *Pfiesteria piscicida* is a toxic dinoflagellate that can cause fish diseases and death in estuarine waters (Steidinger et al. 1996, Burkholder et al. 2001a).

Contaminants in cultures

Ciliates, especially *Euplotes* sp., a herpatrichid ciliate, is a common contaminant of fresh and salt-water rotifer cultures. In general, large populations of ciliates in rotifer cultures have a negative effect on rotifer reproduction (Rotifer Chapter, pages 89, 90 ,91). Yet, *Euplotes vannus* has been found to be beneficial in copepod cultures. Its deliberate introduction into cultures of the copepod *Acartia* resulted in reduced bacterial growth and less accumulation of organic detritus (Zillioux, 1969). This same effect was also thought to occur in rotifer cultures, but our experience suggests that this is not the case. Ciliates are usually found in eutrophic conditions with heavy organic loads and dense phytoplankton (Kinne, 1977).

Reproduction

Under favorable conditions they multiply rapidly by a process called binary fission where they divide in half forming smaller duplicates of themselves. They can also reproduce by conjugation in a similar manner as sexual reproduction in more complex animals.

Paramecium

Paramecium are oval flat creatures, and bear a number of tiny cilia that serve to propel it through the water. As they move through the water they collect small particles of food that are swept into the gullet. Most Paramecia are bacteriovorous and feed voraciously on bacteria that accompany decaying organic matter.

Life Cycle of Ciliates

Euplotes sp., commonly found in saltwater rotifer cultures, are "football-shaped", about one-quarter the length of *Brachionus plicatilis,* with an average width of 20-30 μm and length of 40-50 μm (see Figure 5.11). They seem to be more prevalent in new rather then older well established cultures of rotifers. Reproduction is normally asexual by transverse fission. A form of sexual reproduction also occurs called conjugation where nuclear exchange takes place between two individuals. After nuclear exchange, the cells separate and rapidly divide into four daughter cells.

Physical and Chemical Requirements

Ciliates can be cultured in any size container and have population growth characteristics similar to rotifers. Optimal temperature range for *E. vannus* is 20-25°C (Reguera, 1984). Aeration should be low to moderate. There are some indications that ciliates grow better with a small population of rotifers present. Starter cultures of ciliates can usually be obtained by screening a rotifer culture through a 53-63 μm sieve which retains the rotifers but allows the ciliates to pass through. Some data suggest that ciliates prefer high levels of dissolved organics in the culture water.

Food Requirements

Ciliates, especially *Euplotes,* eat a variety of food from bacteria, algae, yeast, small detritus particles and possibly dissolved organics. Experiments conducted by Reguera (1984) showed that *Euplotes* thrive in cultures of the flagellated green alga *Dunaliella tertiolecta,* in the presence or absence of antibiotics. Food is ingested through a gullet and processed into food vacuoles which are digested as they pass through the cell's cytoplasm towards its posterior region for excretion. Our experiments in culturing ciliates have shown that they are more easily grown on yeast-based foods like Roti-Rich™ or bacteria than on microalgae. Our results on contaminated cultures of rotifers and ciliates indicated that the algae densities did not drop significantly (see Fig 5.10). This indicated that *Euplotes* compete for bacteria and yeast based foods rather than algae.

Experiments by Skogstad, et al. (1987) tested 30 species of freshwater algae for their ability to sustain cultures of five genera of ciliates *(Cinetochilum, Bursaridium, Urotricha, Frontonia, Halteria)*. Their results showed that motile microalgae of the genus Chrysophyceae (2 species), Cryptophyceae (10 species) and Dinophyceae (1 species) were generally good to excellent food. Chlorococcal or flagellate Chlorophyceae (12 species) gave less consistent responses. Diatoms (Bacillariophyceae, 2 species) were excellent for the ciliate *Cinetochilum*, but not acceptable for the others. Blue-green algae (Cyanobacteria, 2 species) were not accepted as food. The ciliate *Cinetochilium,* a small round cell 18-25 μm in diameter, was the easiest to culture and could be concentrated by filtration with a 20 μm screen. As with any cultured plankton, only the concentrated ciliates should be added to larval tanks, not the ciliate culture water.

Culture Method

This is a culture technique used for paramecia but could be modified for ciliates. Place a small tank in a place where it will get some light but is well protected. Make a solution of three packets of activated yeast, two tablespoons flour, and enough water to fill tank well mixed. Place in tank and add culture, an air stone, a good cover, and a heater if needed. Inoculate tank with ciliates.

Once a week strain off half the water from the top of the tank with a coffee maker filter or 10 to 20 micron filter, Feed filtrate and dispose the culture water. Mix a volume of water equal to that taken out with two tablespoons flour and return to tank. Do not add sugar as yeast will "bloom" and contaminate tank. There is plenty of sugar in the waste from the paramecia and the flour to keep the yeast going and the paramecia live just fine off the yeast.

If your dissolved organic protein levels get too high in your culture tank the intermediate sugars from protein decay may spark a yeast "bloom". As long as the water is diluted out before alcohol levels can rise there is no problem. Do a forty percent water change and the culture should be able to control yeast. If the problem persists do another change, and if a third is required it is doubtful your culture can survived in that foul water.Like rotifers, ciliate cultures should be routinely restarted at least every thirty days or less as required.

Additional Points

As indicated by several sources, there may be a synergistic benefit to ciliates by culturing them with rotifers or copepods. If you are having trouble maintaining ciliate cultures, you might want to try a polyculture. In this case, you would want to keep the rotifers and copepods at lower densities and maximize ciliate density. Like rotifers, the nutritional value of ciliates is probably determined by what they eat. Therefore, if ciliates are reared strictly on yeast or Roti-Rich™, a final meal of algae or another lipid rich food is recommended before feeding to larval fish.

Trouble Shooting

1) Too high or too low temperature will greatly reduce ciliate population growth. Most ciliates do not reproduce well at temperatures below 17° C.
2) Ciliate size and cell wall structure appear important in their acceptance and utilization as prey.
3) Polyculture with other zooplankton may have a beneficial effect.
4) Aeration might be too high and strip them from the culture.
5) Optimal food density is not clearly known. Excess food or low concentrations may cause poor growth.
6) Optimal water quality is not clearly known, however, since they are found in rotifer cultures, similar conditions should be maintained.

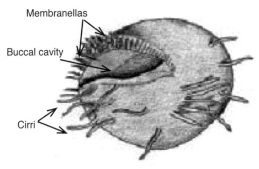

Euplotes
after Coliss (1959)

Chapter 7 - ARTEMIA CULTURE

Introduction

Live Artemia

The brine shrimp *Artemia* has been used as a live food for fish culture since the 1920's. Its widespread use in aquaculture has improved both cyst availability and quality. Brine shrimp cysts are readily available and easily hatched to provide live *Artemia* nauplii for freshwater and marine fish and invertebrates. In addition, more than 700 tons of *Artemia* cysts are sold annually world-wide. Harvesting live *Artemia* and cysts is clearly big business and their use in aquaculture is growing.

Despite recent advances in prepared diets, live foods like *Artemia* and rotifers are still recognized as the best foods for fish and crustaceans. But at a price of $17 to $25 per pound fresh weight, live adult *Artemia* are considered a luxury food by aquarists and far too expensive for many aquaculturists. Recent advances in brine shrimp culture have made it possible to culture adult *Artemia* using simple, reliable procedures and have brought the price of *Artemia* biomass down to more affordable levels (Sorgeloos, et al. 1993).

Live adult *Artemia* provide a high quality, complete protein diet (about 60% protein) which yields better survival, faster growth rates, and fuller color development than most other diets. Live *Artemia* are preferable to frozen because of their higher nutritional value and live foods tend not to degrade water quality. Live *Artemia* nauplii and/or adults are currently used in virtually all commercial shrimp and fish hatcheries. Over 85% of all marine animals now cultured utilize *Artemia* as a partial or sole diet. *Artemia* biomass has also been used as a food additive for domestic livestock and is consumed by humans in Africa and Thailand.

Frozen Artemia

Frozen adult *Artemia* are widely used by aquarists, fish breeders, and aquaculturists. Over 1000 metric tons of adult *Artemia* biomass are harvested annually from natural *Artemia* populations in Canada, France, and the USA. These *Artemia* are quickly frozen and distributed to aquaculture and pet markets. Utilization of live adult *Artemia* is not as familiar to aquaculturists as live nauplii or frozen adults, but the practice is increasing. Recently omega-enriched adult *Artemia* have been introduced to the market. These are harvested, concentrated in large tanks, and fed lipid enrichments like Selco™, squid or herring oil, along with spray-dried microalgae like, Algamac 2000™, and/or *Spirulina*. They are quickly concentrated again, packaged, then frozen.

Life Cycle of Artemia

The brine shrimp *Artemia* is in the phylum Arthropoda, class Crustacea. *Artemia* are closely related to shrimp and zooplankton like copepods and *Daphnia*, which are also used as live foods. *Artemia* life cycles begin by hatching from dormant cysts, which are encysted embryos that are metabolically inactive. Dormancy can persist for several years as long as the cysts are kept dry. When the cysts are placed in seawater, they re-hydrate and the embryos resume development (Persoone, *et al.* 1980; Sorgeloos, *et al.* 1986).

Nauplii development

After 15-20 hr at 25°C, the cyst bursts and the embryo exits the shell (Figure 7.1). For the first few hours, the embryo hangs beneath the cyst shell still enclosed in a hatching membrane (umbrella stage). Inside the hatching membrane the nauplius completes development, its appendages begin to move and it emerges free-swimming. The first larval stage (Instar I) is 400-500 μm in length, weighs about 2 μg dry weight and is brownish-orange in color because of its yolk reserves. Newly hatched nauplii from San Francisco Bay cysts average 430 μm long while those from the Great Salt Lake

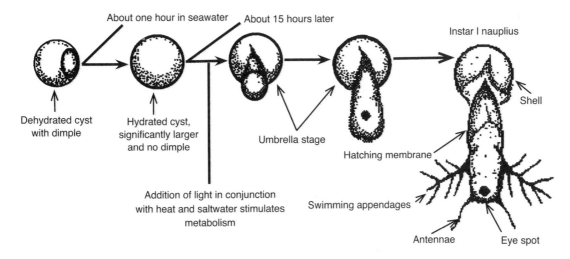

Figure 7.1. Changes of developing Artemia from the cyst stage to Instar I nauplius.

average 489 μm. Newly hatched nauplii (Instar I) do not feed since their mouth and anus are not developed. Approximately 12 hr after hatch, they molt into the second larval stage (Instar II) at which time feeding commences. Instar II nauplii begin filter-feeding on 1-40 μm particles, including various microalgae, bacteria, and detritus. Nauplii grow and progress through 15 molts before reaching adulthood at 8 days. The new nauplius has only 3 pairs of legs and new ones are added with each of the first 10 molts.

Adult development

Adult *Artemia* average 8 mm long in bisexual populations, but can attain lengths of 20 mm in polyploid parthenogenetic populations. Adults typically weigh about 1000 μg dry weight, which represents a 20-fold increase in length and a 500-fold increase in biomass from the naupliar stage (Reeve, 1963). *Artemia* adults are non-selective filter-feeders which consume microalgae, bacteria, and detritus. Adults can live up to four months if maintained in good conditions. Males are easily distinguished by the muscular graspers which are modified antennae in their head region. These are used to cling to the female for many hours while breeding. Females are distinguished by the brood pouch at their posterior (Figure 7.2).

Cyst Production

At low salinities and optimal food levels, fertilized females usually produce free swimming nauplii (ovoviviparous reproduction) at a rate of up to 75 nauplii per day. They may produce 10-11 broods over an average life cycle of 50 days. Under ideal conditions adult *Artemia* survive for several months and produce up to 300 nauplii or cysts every 4 days.

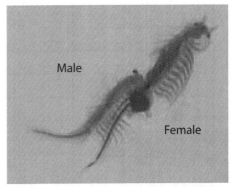

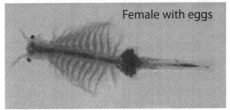

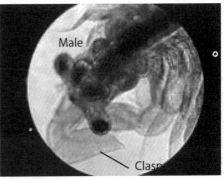

Figure 7.2. Morphology of adult male and female Artemia.. Photos by P. Gonzalez, G. Ramirez & Y. Briceno

Cyst production (oviparous reproduction) is induced by high salinity, under conditions of high eutrophication (large O_2 fluctuations between day and night), and chronic food shortages. At high salinities (>150 ppt) and low oxygen concentrations, shell glands accumulate a brown compound called hematine. Embryos develop to the gastrula stage, become surrounded by a thick shell and enter dormancy. Under these conditions cyst production (oviparous) becomes dominant. Females can release up to 75 cysts per day which float in the high salinity water. The floating cysts are eventually blown ashore where they accumulate in large masses and dry. Development is resumed when the cysts are re-hydrated and the life cycle is begun again.

Distribution

Strain	Temperature (°C)	Salinity Range (ppt)
San Francisco Bay USA	21-24	15-120
Macau, Brazil	21-26	35-115
Great Salt Lake USA	26-31	15-120
Manaure Columbia	23-33	not known
Galera Zamba, Columbia	21-31	not known
Chaplin Lake, Canada	21-28	40-75
Buenos Aires, Argentina	21-26	15-120
Lamaca, Cyprus	21-26	15-120
Barbanera, Spain	21-28	15-110
Shark Bay, Australia	23-28	not known
Tuitorin, India	23-31	not known
Tienntsin, China	23-31	not known
Margherita di Savoia, Italy	23-28	not known

Table 7.1 - Comparison of temperature and salinity optima for Artemia strains from different geographical regions (Sorgeloos, et al 1986)

Ecologically, Artemia are distributed all over the world. Cysts are dispersed by seagulls, ducks and flamingos, and can pass through the digestive tract of birds unharmed. As a cosmopolitan species, Artemia is adapted to a large range of environments (Table 7.1).

Brine shrimp grow at temperatures ranging from 6-35°C (43-95°F) and in waters of greatly different ionic composition (chlorides, sulfates, and carbonates). Artemia can thrive in open seas and coastal waters, but are quickly consumed by fish, crustacean and insect predators. Since Artemia are equipped with few predator defenses, their distribution has become restricted to harsh environments which are too extreme for predators. Waters with salinity of 70 ppt or higher are the most likely habitats to find natural populations of Artemia. Although Artemia is physiologically capable of surviving in brine up to 340 ppt, the optimal salinity for adult growth and oviparous reproduction is about 30 ppt.

Physical and Chemical Requirements

Artemia tolerate a wider range of environments than most animals, but grow best in specific conditions. Several environmental parameters which affect Artemia survival and growth are outlined below. The following summary of data is from Sorgeloos, et al. (1986) where it is discussed in greater detail.

Temperature

Adults tolerate brief exposures to temperatures as extreme as -18° to 40°C (0-104°F). Optimal temperature for cyst hatching and adult grow out is 25-30°C (77-86°F), but there are differences between strains. For example, the temperature optimum for the San Francisco Bay strain is 22°C (72°F) as compared to 30°C (86°F) for the Great Salt Lake strain. For most strains, constant rather than fluctuating temperatures seem best. Dry cysts can withstand even lower and higher ranges -273° to 60°C (-523-140°F) without affecting viability. Long term storage of cysts should be in cool, dark, low humidity places, preferably refrigerators or freezers. Open cans of cysts are highly vulnerable to decay due to moisture absorption and should be tightly covered, kept cool, and stored at low humidity.

Salinity

Artemia are excellent osmoregulators, maintaining the salt concentration of their body tissues within narrow limits of 9 ppt even when living in brine ponds of 180 ppt salinity. Although *Artemia* tolerate high salinities, they prefer 30-35 ppt (1.0222-1.0260). *Artemia* nauplii and adults survive in freshwater for about 5 hr before they stop swimming, sink to the bottom, and die. Caution must be exercised when using *Artemia* in freshwater, since overfeeding can lead to deterioration of water quality as animals quickly decompose. However, many freshwater fish and invertebrates tolerate low salinities of 1-5 ppt easily, so it is possible to add seasalts to freshwater and extend survival of *Artemia* .

Water composition

Artemia tolerate water with a wide range of ionic composition. The following are examples of the ionic ratios tolerated as compared to natural seawater: sodium/potassium= 8-173, seawater 28; chlorine/carbonate= 101-810, seawater 137; and chlorine/sulfate= 0.5-90, seawater 7. In spite of this wide tolerance, the ionic composition of natural seawater is still preferred. The ionic composition of some salts, like rock salt or water softener salts, may be beyond the range tolerated by *Artemia*. Best cyst hatching and adult grow-out results will be obtained when the ionic composition of the culture medium closely approximates seawater. In addition, like *Daphnia*, *Artemia* tolerate high levels of ammonia up to 90 ppm (Sorgeloos, 1980).

pH and light

Other environmental variables of importance are pH, light, and oxygen concentration. A pH of 8 to 9 is best; pH less than 5 or greater than 10 is usually lethal. The pH of seawater may be increased with sodium bicarbonate (baking soda) and lowered with hydrochloric acid (muriatic acid). A minimum amount of light is necessary for hatching and is also probably beneficial for adult grow out. An intensity of about 2000 lux with a spectral quality approximating sunlight is adequate.

Oxygen

The level of oxygen actually dictates what *Artemia* consume. At favorable oxygen levels the animals are a pale pink or yellow or, if feeding heavily on microalgae, they will be green. Under these ideal conditions growth and live nauplii reproduction is rapid. When the environment has low oxygen levels and high amounts of organic matter, or a high salinity, they feed on bacteria, detritus, and yeast cells, but little algae. It is under these conditions that they produce hemoglobin and are red or orange in color. If these conditions persist, *Artemia* become oviparous and produce cysts (Ivleva, 1969). Oxygen levels should always exceed 2 ppm (mg/L). Heavy aeration or high surface area and a shallow depth usually provide sufficient oxygenation.

Food Requirements

Food values

New hatched *Artemia* (<0.6 mm in length) are high in fats with a range of 12-32% of the dry weight (Dutrieu, 1960). In the metanauplius stage (2.5 mm) the fat levels decrease to 16.5% and by the time the nauplii reach a pre-adult stage fat levels decrease to 7%. The protein content increases from 42.5% in new hatched nauplii to 62.8% in adult stages (Helfrich, 1973). Based on this knowledge, culturists should determine at what age brine shrimp should be harvested and fed to obtain the best food values. For example, young fish larvae need higher fat levels while older juveniles need more protein. Letting *Artemia* grow to adults before utilizing them as food may not be advantageous and is more costly.

Table 7.2 is a combined biochemical and nutritional analysis of *Artemia salina* nauplii hatched from cysts from various sources (Creswell 1993). As shown there is a wide variability of nutrients from different source of the same specie. Fatty acids values with an asterisk may be totally absent from certain sources.

Like rotifers, the food value of *Artemia* can be improved by enrichment with highly unsaturated fatty acids. The procedure for enrichment is similar to that described in the Rotifer Chapter. A detailed description of *Artemia* enrichment techniques and an evaluation of commercially available products is presented by Sorgeloos, et al. (1993). Techniques for enrichment of rotifers and *Artemia* with therapeutic compounds like antimicrobial drugs are described by Verpraet, et al. (1992).

Types of feed

Since they are non-selective filter feeders, a wide range of living and inert foods have been used successfully to culture brine shrimp. Criteria for food selection is based on particle size (<50 to 60 μm), digestibility, and solubility. Feeds with high solubility should be pre-soaked or avoided. Microalgae that have been used as feeds include *Nannnochloropsis, Dunaliella, Chaetoceros, Phaeodactylum, Tetraselmis*, and *Isochrysis*. A wide variety of inert foods are also acceptable, but as with any inert food, water quality problems can easily develop from over feeding. Inert feeds include active and inactive yeast, micronized rice bran, whey, wheat flour, soybean powder, fish meal, egg yolk, and homogenized liver. Dried microalgae such as *Spirulina, Scenedesmus* and *Tetraselmis* have also been used successfully. Bacteria growing on dried foods contribute significantly to their nutritional value (Doulliet, 1987). Some types of bacteria cells can themselves serve as an effective food source for *Artemia* (Intriago and Jones, 1993).

Biological Composition	Measurement
Individual Dry Weight (mg)	1.48
Ash Weight (% dry weight)	11.28
Total Lipid (% dry weight)	13.7
Fatty Acids (% dry weight)	10.9
Coloric contact/gm Ash free (cal)	5.503
Individual Coloric Content (μcl)	7.30
Water (% wet weight)	90.85
Dry Matter (% body weight)	9.15
Carbon (% body weight)	27.5
Nitrogen (% body weight)	8.09
Phosphorus (% body weight)	1.24
Essential Amino Acids	
Threonine (g/100 gram protein)	4.8 - 6.0
Valine (g/100 gram protein)	3.1 - 5.5
Methionine (g/100 gram protein)	2.2 - 3.7
Isoleucine (g/100 gram protein)	4.9 - 6.8
Leucine (g/100 gram protein)	7.9 - 10.1
Phenylalanine (g/100 gram protein)	5.1 - 10.4
Histidine (g/100 gram protein)	2.7 - 4.9
Lysine (g/100 gram protein)	8.7 - 11.7
Arginine (g/100 gram protein)	9.7 - 11.5
Fatty Acids (Dry Weight)	
18:0 Stearate (mg/gram)	2.79 - 6.83
18:1Ω9	26.97 - 31.2
18:2Ω6 Linoleate	3.69 - 9.59
18:3Ω3 Linolenate	4.87 - 33.59
18:4Ω3	0.96 - 4.88
20:1Ω9 Eicosaenoate	0.35 - 0.52
20:2Ω6/9	0.06 - 0.24*
20:3Ω6 Eicosatetraenoate	0.05 - 2.76
20:3Ω3/20:4Ω6	1.48 - 2.69*
20:5Ω3	1.68 - 13.63
22:6Ω3 Docosahexaenoic	0.06 - 0.26*

Table 7.2 - Nutritional values of various sources of Artemia salina new hatched nauplii.

Food levels

Perhaps the simplest way to measure food level is by estimating water transparency. This can be done with a density measuring stick, which is a vertical scale with a white disk attached to its end. The depth where the white disk just disappears measures light penetration into the medium. Higher concentrations of food in the medium reduce transparency. With a stocking density of 5000 nauplii per liter, transparency should be 15-20 cm the first week and 20-25 cm thereafter. It is best to maintain food near optimal levels, so frequent feedings or continuous drip are desirable. Once per day is probably a minimum. However, nauplii do not commence feeding until they are about 12 hours old so initial feeding can be delayed, especially if using a non-living food.

Mechanism of feeding

Food is not directly consumed, but rather transferred to the mouth in packed form. The space between the legs widens as the legs move forward on the forestroke. Water is sucked into this space from the area below the midline of the body. Small filtering setae (bristles) collect particles including food items, from the incoming stream. On the backstroke, water is forced out through the space into between the legs, but the concentrated food particles remain in a food groove below the base of the legs. This food groove channels the food towards the mouth. Trapped food particles are transferred to the mouth by a complex mechanism. Glands along the groove secrete adhesive material that clumps the particles into food balls. When feeding on microalgae, *Artemia* may turn a green color (Hunter, 1981).

Procedure For Hatching Cysts

Hatching requirements

Optimal environmental conditions for hatching *Artemia* cysts are temperatures of 25°-28°C and salinities of 15-35 ppt. Seawater can be made from commercial seasalt mixtures like Forty Fathoms™ or Instant Ocean™. Oxygen should be saturated (continuous heavy aeration), light, about 2000 lux constant illumination, and pH, 8-9. Good circulation is essential to keep the cysts in suspension. Containers that are V-shaped, inverted pyramids or cylinders with steep concave bottoms provide the best circulation patterns and are convenient for nauplii collection. When these containers are equipped with bottom valves or drains, unhatched cysts, hatched nauplii and empty shells can be easily removed (Figure 7.3). Hatching percentage and hatching density are usually a function of circulation patterns and water quality. Containers with dead areas such as garbage cans and flat bottom containers are not good for maximum hatch.

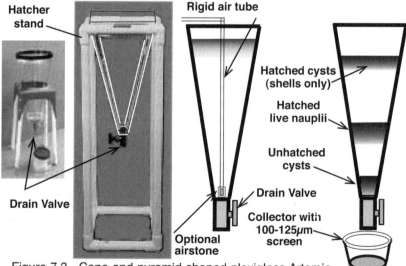

Figure 7.3. Cone and pyramid shaped plexiglass Artemia hatchers made by Florida Aqua Farms.

Hatching density

Cyst producers suggest an initial density of about 1.65 grams of cysts per liter (0.5 teaspoon). Using the proper shaped hatcher like the 17 liter pyramid shaped hatcher shown in Figure 7.3, we have increased the density as high as 8.9 grams (270 teaspoons) per liter. However, 5.5 grams (1.2 teaspoons) is a good density for consistent results. Depending on quality, size, and weight of cysts, usually 200,000 to 300,000 nauplii hatch per gram of cysts and these are about 3.3 grams per level teaspoon or 4.93 ml. Biomass of 30 ml (1 fl. oz.) of *Artemia* cysts (grade A) yields about 47 ml (1.6 fl. oz.) of hatched *Artemia* nauplii.

Timing hatch

Hatching can vary with the age of the cysts, where they came from, and whether optimal hatching conditions are maintained. Older cysts usually take longer to hatch and the percentage hatched is lower. Once a can is opened, moisture absorption is a serious problem that can drastically affect hatch regardless of cyst age. Cysts should be divided into smaller jars and stored in a refrigerator or freezer. New cysts begin hatching after 18 to 24 hours. By 36 to 48 hours most cysts will have hatched. Often a two stage harvest is used where at 18-24 hours the initial harvest is made followed by another at 36-48 hours. There is always a percentage of cysts that do not hatch, yet these are valuable as a food source if decapsulated (removal of the external shell). It is also possible to increase the percentage hatched by decapsulation.

Harvesting nauplii

Harvest nauplii by turning off the aeration and letting the culture settle for about 10 minutes. Hatched, empty shells float on the surface and unhatched cysts sink to the bottom. Newly hatched nauplii will concentrate just above the unhatched cysts on the bottom (see Figure 7.3). Since the nauplii are positively phototropic (attracted to light), shining a light at the middle of the container helps direct them to an area where they can be easily harvested by siphoning or draining. Shading the container at the top and bottom will exaggerate this effect. Clean nauplii can then be concentrated with a zooplankton collector. The concentrated nauplii can be washed and added directly to the culture tank. When using a bottom drain, draw off the unhatched cysts into a separate container for future hatching before collecting the nauplii. Nauplii tolerate direct transfer to salinities of up to 150 ppt without ill effects. As when feeding live rotifers, only the concentrated *Artemia* should be added to the larval tank, not the *Artemia* culture water.

Procedure For Decapsulating Cysts

Reasons for decapsulation

Decapsulation to separate nauplii from their shells is desirable for several reasons. Cyst shells are indigestible and can lodge in the gut of predators, causing obstructions that ultimately can be fatal. Shells can be contaminated with bacteria and sources of infection. Decapsulation disinfects the cysts, facilitates removal of empty shells, increases cyst hatching percentage, and provides a 100% edible product regardless of hatch rate. Cyst decapsulation or light chlorination has therefore become a standard procedure in many commercial hatcheries Sorgeloos, et al. (1993).

Cyst hydration

Decapsulation is the removal of the hard, dark brown external layer of *Artemia* cysts called the chorion. The chorion can be dissolved by brief treatment with a hypochlorite solution without harming the embryo inside (Sorgeloos, et al. 1977). Decapsulation is accomplished in four steps: hydration of cysts, treatment with decapsulating solution, washing and deactivation of residual chlorine, and hatching or storage of the embryos. These steps are described in detail below.

Dry cysts have a concave surface or "dimple" which makes it harder to remove the complete chorion. Therefore, the cysts are first hydrated into a spherical shape. Cysts should be hydrated in fresh or salt water (1 gram of cysts /30 ml water) at 25°C for 60-90 min. Hydration takes longer at lower temperatures, but should not exceed two hours since prolonged hydration decreases hatching rate and hatching percentage of decapsulated cysts. Hydration should be done in a conical or V-shaped container which maximizes circulation of cysts with moderate aeration without an airstone. Cysts should be filtered on a 100- 125 μm collection screen and rinsed.

Decapsulation process

It is best to decapsulate hydrated cysts immediately, but if necessary they can be stored in a refrigerator at 4°C for several hours. The chlorine decapsulation solution should be mixed with saltwater during the hydrating process. Either fresh household liquid bleach (5.25% sodium hypochlorite = NaOCl) or powdered pool chlorine (calcium hypochlorite = CaOCl) can be used.

In preparation for decapsulation, cysts are placed in a pre-cooled buffered solution (4°C and approximately pH 10) consisting of 0.3 ml of 40% sodium hydroxide (NaOH) and 4.7 ml of seawater per gram of cysts. The buffer stock solution is prepared by dissolving 40 g NaOH in 60 ml of freshwater. Decapsulation begins when 10 ml of liquid bleach per gram of cysts (1.65 gm = 0.5 teaspoon) is added to the buffed saltwater solution. Since heat is given off during decapsulation, it is important to maintain the buffer solution temperature between 20-35°C. Starting with pre-cooled buffered seawater makes it easier to keep the reaction in this temperature range. The saltwater mixture can be made with common water softner rock salts at 18-30 ppt. (1.0130 to 1.0222 density), but if poor hatching occurs, use commercial seasalts. Ice can also be added to reduce temperature during the reaction.

A second method is to add 0.70 g of dry pool chlorine powder per gram of cysts. In this case the buffer is sodium carbonate (Na_2CO_3) consisting of 0.68 g Na_2CO_3 in 13.5 ml water per gram of cysts. It is easier to split the required pre-cooled saltwater into two equal parts. Add the required amount of powdered chlorine to the first part and the Na_2CO_3 to the second part. Allow them to dissolve and react which will cause a precipitate. Pour off the clear portions (supernatant) and discard the precipitates. Mix the two solutions together and add the hydrated cysts.

During decapsulation stir and aerate continuously to minimize foam formation as the chorion dissolves and to dissipate heat. Note the color change. With liquid bleach the color of the solution will change from a dark brown to grey, to white, and then to a bright orange. This reaction usually takes 2-4 min. With calcium hypochlorite the cysts will change only to gray and will take slightly longer (4-7 min).

The cysts can be filtered from the solution as soon as the chorions have dissolved as indicated by color. The chlorine should be washed off the cysts by rinsing with fresh or saltwater until no chlorine smell is detected. Residual chlorine attaches to decapsulated eggs and must be neutralized. Washing the cysts in a 0.1% sodium thiosulfate (0.1 g sodium thiosulfate in 99.9 ml water) for one minute is the best method. An alternative method uses acetic acid (1 part of 5% vinegar to 7 parts water). The cysts are then re-washed with fresh or saltwater before placing them into hatching medium. Decapsulated cysts can be hatched immediately or stored at 4°C for up to 7 days before hatching. A weekly supply can be made once each week.

Storage of decapsulated cysts

For long-term storage, decapsulated cysts must be dehydrated. Dehydration of decapsulated cysts is accomplished by transferring 1g of decapsulated cysts into 10 ml of saturated brine (330 g NaCl/l) and aerating for 18 hr. Due to the osmotic imbalance, cysts release water to the brine solution which should be replaced every 1-2 hr or more salt added. After 18 hr, cysts have lost about 80% of their cellular water. Aeration can be stopped, the cysts settled and collected by filtration. The dehydrated, decapsulated cysts can now be transferred to a container, topped off with fresh brine, sealed and stored in the refrigerator or freezer. Decapsulated cysts with 16-20% cellular water content can be stored for a few months without decrease in hatching rate or percentage. Longer storage requires cellular water to be reduced to less than 10%. To hatch stored cysts the hatching protocol described above should be followed.

Cold Storage of Nauplii

Once *Artemia* cysts hatch, nauplii grow rapidly on stored energy reserves, reaching Instar II in several hours. Older (larger) nauplii may be too big for larval fish to handle and are more difficult to catch. As newly hatched Instar I nauplii grow to Instar II, they lose 22-39% of their energy content (Sorgeloos, et al. 1993). Consequently, it is best to use the nauplii immediately. If the nauplii are not needed right away, it is possible to slow down naupliiar development by storage at <10°C. Nauplii can be concentrated to densities of 5 million/L for periods of up to 24 h without deleterious effects (Leger, et al. 1983). Mild aeration may be necessary to prevent condensed nauplii from accumulating on the bottom and suffocating. Cold storage of nauplii allows for large daily batches to be hatched and used throughout the day thus saving time and labor. Continuous availability of nauplii allows for more frequent larval feedings and more efficient food uptake.

Procedure For Production of Adults

Changes in nutritional values

When feeding larger fish and invertebrates where small food particle size is not required, *Artemia* adults are preferred over nauplii. Adult *Artemia* are 20X longer and 500X heavier than nauplii and therefore provide much more meat per bite. The objective of culturing *Artemia* to adulthood is to multiply the biomass of a small amount of cysts into a large amount of adults. It is possible, for example, to convert 10 g of cysts into 2000 g of adult *Artemia* in 2 weeks of intensive culture.

If nauplii are not fed 12 hours after hatching, they rapidly lose weight and caloric value. Unfed Instar II nauplii lose 20% of their weight and 27% of their caloric value by the time they molt to Instar III (about 24 hr). Starving nauplii have little nutritional value.

If properly fed, favorable compositional changes also occur as *Artemia* grow from nauplii to adults. The dry weight of nauplii is 20% lipid and 42% protein as compared to 10% lipid and 60% protein in adults. Nauplii are known to be deficient in the amino acids histidine, methionine, phenylalanine and threonine, while adults are rich in all essential amino acids. Adult *Artemia* therefore supply more biomass than nauplii and are more nutritionally complete.

Culture tank design

There are several approaches to growing *Artemia* adults in commercial aquaculture. The batch culture technique is simplest and described here. Best yields are obtained with good food circulation, animal distribution, and moderately strong aeration. Rectangular aquaria or common plastic storage containers can be easily modified into small raceways, enhancing circulation patterns and improving production.

This is accomplished by mounting a partition down the middle that is positioned equal distance from all sides (Figure 7.4). To further enhance circulation, the ends of the aquaria can be rounded by gluing curved pieces of formica or plexiglas in the corners with silicone glue or "Goop" depending on the type of media being glued. Height/width ratio of the culture is important which, according to Sorgeloos, et al. (1986), should be smaller than 1. Good circulation patterns are created by mounting 4 or more air/water lifts on the sides of the partitions (Figure 7.4). Optimal circulation and aeration is insured when pipes are spaced at 25-40 cm intervals (10-16"). Pipe diameter of the lifts is dependent on water depth. At a depth of 200 mm (8"), a 25 mm (1") diameter pipe is best, at 400 mm (16") depth - 40 mm pipe (1.5"), 750 mm (30") depth - 50 mm (2") pipe, and at 1000 mm (40") depth - 65 mm (2.5") pipe.

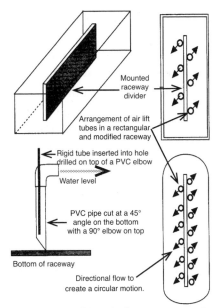

Figure 7.4. Circulation pattern for a closed raceway culture tank for Artemia (modified from Sorgeloos et al. 1986).

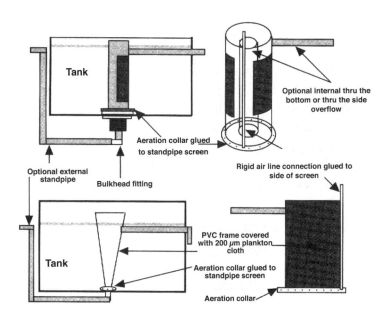

Figure 7.5. Overflow screen designs for flow-through Artemia culture systems (modified from Sorgeloos, et al. 1986).

Aeration

Like rotifers, *Artemia* cannot tolerate fine bubbles since they can lodge in the swimming appendages and cause mortality. *Artemia* have been noted to actually ingest small air bubbles which incapacitate them. Therefore, it is best to aerate without airstones or use those which create coarse bubble stones.

Closed system filtration

This closed raceway can be further modified into a continuous filtration or water exchange system by providing a screened overflow (Figure 7.5). This can be constructed from a PVC pipe or side screen mounted in the tank and covered with plankton screening. To help reduce clogging, an air collar made of porous rubber or punctured vinyl tubing can be glued at the base of these collecting screens. As the shrimp grow, larger mesh screens are sequentially put in place starting with 200 μm and progressing to 250, 300 and 400 μm. Further filtration can be added by installing a sump or reservoir, with biological substrata, to receive primary effluent from the culture tanks, followed by cartridge or sand mechanical filtration and a UV. For more information about large scale batch and continuous cultures see Sorgeloos, et al. (1986) or Dhert, et al. (1992).

Food and feeding

To initiate a high intensity batch culture, freshly hatched nauplii are collected and placed in a 5 or 10 gallon aquarium at a stocking density of 1000-3000 per liter (3750-11250/gal). Newly hatched nauplii do not feed for approximately the first 12 hours. Therefore, feeding of algae can begin the next day. Optimal algal cell density varies with the algal species used because different species have different cell sizes and reproductive rates. When using *Chaetoceros*, for example, optimal cell density is 50,000 cells/ml. In contrast, *Nannochloropsis* cells are about 50% smaller, so its density should be maintained at about 100,000 cells/ml. Because of this variation in cell size, the most convenient way to estimate food level is to measure water transparency with a Density Measuring Stick™ (available from Florida Aqua Farms) or turbidity meter.

Physical Culture Parameters

Since Artemia feed constantly, faster growth rates and higher survival are achieved by multiple or continuous feedings over a 24-hour period. Better growth rates are achieved at 25-30° C, salinities of 30-50 ppt, and low light levels. These are optimal growth conditions for most strains, however, strains from different geographical regions may differ substantially in their temperature and salinity optima (Table 7.1). Some strains have narrow temperature ranges like Tuticorin, and others have broad ranges like Manaure. Similarly, several strains tolerate a wide range of salinity, but some, like the Chaplin Lake strain, have a considerably narrower optimum range. *Artemia* are drawn towards strong light, which causes increased swimming activity and greater energy expenditure, resulting in slower growth rates. In low light, *Artemia* spread out in the water column, swimming slowly and achieving more efficient food conversion.

Artemia growth rates and the quantity of animals reared are influenced by several factors. Compared to rotifers, *Artemia* are more sensitive to low oxygen concentrations, high levels of dissolved nutrients, bottom detritus, and starvation. Moderate aeration without airstones, good water quality, and generally clean conditions are all important for raising high densities of adult brine shrimp.

Maintenance

Maintenance problems increase as the density and biomass of a culture increases. Water quality rapidly deteriorates in small volumes of water and maintenance must be rigidly controlled. The primary problem is usually due to over feeding, leading to fouling followed by low oxygen concentrations and eventual mortality. There is a fine line between optimal feeding level and degradation of water quality, especially when using non-living foods. Fecal wastes and flocculate particles accumulate rapidly and can actually physically hamper movement and ingestion of feed particles. To overcome water quality problems, commercial high density systems often include flow-through (open system), routine partial water exchanges (semi-open system), or continuous closed water exchanges (closed recirculating systems). Closed culture systems include particle filters, protein skimmers, and biological filtration (van Rijn 1996). In addition, utilization of live microalgae rather than inert feeds helps minimize maintenance, especially in static batch cultures.

Maintenance should include bottom cleaning and/or flow-through removal at least every day. Cleaning can be accomplished by turning off the air, providing a strong light at the surface to attract the brine shrimp then siphon detritus and fecal matter off the bottom. If the tank is not part of a recirculating water system, at least 20% water exchanges should be made each week. Water exchanges can be incorporated into the daily additions of microalgae if this is used as the primary food source. Foaming may occur especially if you are using yeast, soy bean, rice bran, or other types of micronized foods. This can be controlled by adjusting air flow and food levels.

Procedure for batch cultures

The following is a protocol developed by Sorgeloos, et al. (1986) for batch culture of adult *Artemia* using the modified raceways described in Figure 7.4. Typical survival rates, growth, and biomass production using this system are illustrated in Figure 7.6. Note the significant mortality between day 4 through 6 and significant biomass increase and growth between days 8 to 12.

1. Natural or synthetic seawater at 30-50 ppt, pH>8.0 (if lower add 1 g/L NaHCO₃), and maintain temperature at 25-30° C.

2. Harvest newly hatched nauplii. Wash on a 100 μm screen with fresh seawater before transferring into the culture tank. Stock at a density of 1000-3000/L (Note: stocking densities as high as 5000/L can be used as you gain experience).

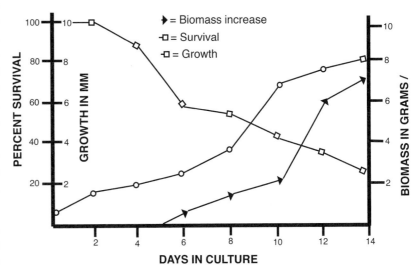

Figure 7.6. Survival, growth rates, and biomass production of Artemia in a 2 cubic meter raceway at 25°C, 35 ppt, and stocked at 5000 animals per liter (Sorgeloos, et al. 1986).

3. Feed should not be offered for at least 12 hours after stocking.

4. Feed can be administered manually or by timer. Transparency measurements of culture water indicate food levels and can be measured with the Density Measuring Stick™. During the first week of culture, maintain water transparency at 15-20 cm and 20-25 cm thereafter.

5. As the animals grow, the daily ratios of feed will have to be adjusted by either increasing the quantity of each feeding or by increasing feeding intervals.

6. Beginning on the 4ᵗʰ day, waste materials should be siphoned from the bottom or removed by turning on the flow and forcing the waste particles through a 200 μm screened overflow tube into a central filter.

7. If using a continuous overflow tube it should be cleaned at least every other day and the mesh size increased as the animals grow.

8. pH and oxygen levels should be monitored. When the pH falls to 7.5, adjust to 8.0 with sodium bicarbonate (about 0.3 g/L). When oxygen decreases below 2 mg/L increase the aeration or change half the culture medium with fresh seawater.

9. The health status of the animals can be verified by their swimming activity. The quick concentration of animals when exposed to strong light is indicative of good health. However, slow, dispersed swimming activity indicates stress. Another indicator is microscopic inspection of the digestive track which should be packed with food. The swimming appendages and mouth region should be clean; if covered with food particles the animals probably are starving. This condition may be due to the nature of the feed or the physiological condition of the animals.

10. Monitoring animal density gives a good indication of culture productivity. Density measurements can be made by removing a measured volume of water, counting the animals and multiplying this by the volume of the tank. When counting animals, note their size and variability, which are indicators of feeding efficiency, food availability, and distribution.

11. The culture duration varies with temperature, feeding regime, and the *Artemia* strain selected. Usually *Artemia* will reach adult size in about 2 weeks at 20-25°C.

12. After harvest, disinfect the culture tank before starting a new culture.

Using this batch culture technique in culture containers of 300 to 5000 liters (80 to 1333 gal) with inert micronized foods, Sorgeloos *et al.* (1986) achieved an average of 5-7 kg (11 to 15 lbs) wet-weight *Artemia* biomass/m^3. Average survival was about 30% and the biomass increased 7X over the culture period. Utilization of flow-through systems and more refined feeding allowed for stocking densities as high as 20,000 nauplii/liter.

Utilization of harvested Artemia

Adult *Artemia* are best harvested by stopping aeration for about 20-30 minutes to allow debris to settle. Adults congregate near the surface where they can be siphoned off into a collector, rinsed, and added directly into the fish tank. Remember, if *Artemia* are being fed to freshwater organisms, they should be rinsed with freshwater before placing them into the freshwater tank.

If more *Artemia* are being produced than can be immediately used, store the biomass for future use. Adult *Artemia* survive for several days in a refrigerator (see rotifer biomass storage). If adults have been refrigerated for several days, warm them up and feed them one last time before placing them in the fish tank. This last feeding restores most of their nutritional quality after several days of starvation. *Artemia* biomass can also be frozen (freeze in 7-8 ppt saltwater) for use later. Ice cube trays are useful for this purpose. The *Artemia* biomass should be drained and frozen as quickly as possible for the best results. Frozen *Artemia* cubes can be placed directly into fish tanks and will slowly thaw, releasing intact brine shrimp. Freezing kills *Artemia*, so caution must be used not to overfeed. Dead *Artemia* settle from the water column and decompose like any food, which can lead to water quality problems.

Trouble Shooting

Possible reasons for slow growth are listed below:

1) Insufficient aeration causing asphyxiation - increase aeration levels, but avoid small bubbles.

2) Low food levels causing starvation - increase food level without degrading water quality.

3) Mortality, slow growth or slow swimming, poor water quality resulting from accumulation of dissolved organics, nitrogenous wastes like ammonia, or bacterial toxins. Increase water exchanges and removal of accumulated feces and flocculate particles. Decrease inert feed levels and increase microalgae levels.

4) Slow growth - low temperature, pH imbalance, too high or low salinity, inadequate food, or poor food quality.

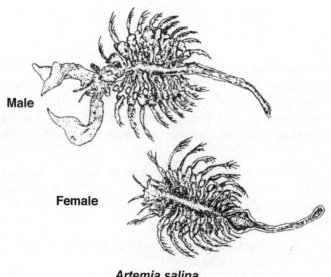

Male

Female

Artemia salina

Chapter 8 - COPEPOD CULTURE

Introduction

Crustaceans is the largest phylum (10,000+ species) within is Copepoda are the largest class of crustaceans. Copepods form and important link between phytoplankton and higher trophic levels in most aquatic ecosystems (Lavens and Sorgeloos,1996). They are found everywhere and are significant prey for a variety of fish and invertebrates. They occur in marine and freshwater environments but most are marine. Many are parasitic, others swim freely (planktonic), while others live in benthic (bottom dwelling) communities. Few planktonic varieties exceed 2 mm in length. Of the three free-living (non-parasitic) orders, the calanoids are largely planktonic, the harpacticoids are largely benthic, and the cyclopoids contain both planktonic and benthic species (Barnes, 1963).

Copepods are important natural zooplankton communities. Since they belong to the Class Crustacea and like others in this class, their body is typically divided into a head, thorax, and abdomen. They are further placed in the Subclass Copepoda which is divided into eight Orders: Calanoida, Cyclopoida, Harpacticoida, Monstrilloida, Notodelphyoida, Caligoida, Lernaeopodoida, and Arguloida. In nature, copepods bloom seasonally, forming enormous clouds which drift in the surface waters of bays and open oceans.

There are about 2300 described species of calanoid copepods worldwide, of which some 25% occur in freshwater. Approximately 98 species occur in North America and 15 of those may occur in the Great Lakes. It has been suggested that two of these 15 species are the result of canal openings (Eurytemora affinis) and bait bucket introductions (Skistodiaptomus pallidus). The Great Lakes species include representatives of four (Aetideidae, Centropagidae, Diaptomidae, Temoridae) of the 40 known families of calanoids.

There are pelagic, benthic, and parasitic forms, so care must be taken in selecting one that will suit your needs. The ease of culturing is certainly not the only criterion for selection of copepods since some species are carnivorous predators and can kill fish and invertebrate larvae. Species of the orders Notodelphyoida, Monstrilloida, Caligoida, and Lernaeopodoida are exclusively parasitic. In most cases, aquaculturists culture pelagic or benthic species of the orders Calanoida, Cyclopoida, and Harpacticoida. More desirable Calanoid species can be distinguished by their long first antennae with 16 to 26 segments, while harpacticoids have short first antennae with fewer than 10 segments.

In recent years there has been increased interest in finding alternative live foods which can be used

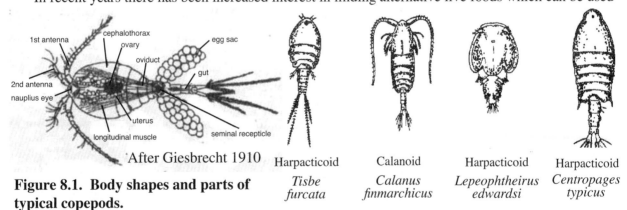

Figure 8.1. Body shapes and parts of typical copepods.

After Giesbrecht 1910

| Harpacticoid | Calanoid | Harpacticoid | Harpacticoid |
| *Tisbe furcata* | *Calanus finmarchicus* | *Lepeophtheirus edwardsi* | *Centropages typicus* |

exclusively or in conjunction with other live foods (Stottrup, 2000). Ogle et al. (in press) is considered one of the pioneers in copepod culture in the USA. He and his associates summarized the current status of copepod culture. Although there is a recognized need, copepods are not used routinely in commercial aquaculture because compared to brine shrimp and rotifers they are difficult to culture in sufficient quantities in a timely manner is problematic. Culture of copepods is developing along several lines from extensive culture to intensive culture. On the extensive end lie production methods that produce low densities of mixed species typically in ponds that are fertilized to provide blooms of food for copepods. Typically fish larvae are cultured in the same pond as the bloomed copepods, a combination referred to as "endogenous" production. On the intensive end lie methods that produce high densities of single species in a controlled environment using separately cultured sources of nutrition that are supplied to the copepods in a prescribed feeding regimen. Typically fish larvae are cultured separately from the copepods and the partitioning of copepod production from the rearing of larvae is referred to as "exogenous" production. For the most part, as culture intensifies both the density of copepods increases and the effort required to operate the system increases. The appropriate choice is dependent on the resources and goals of a facility and should optimize he yield per unit of effort.

Efforts to culture copepods in low densities is not overly hard but utilize routine methods to reach success. Research and hobby level needs can be usually satisfied. To date most culture efforts have been concentrated on the calanoid *Acartia tonsa*. In extensive production it usually becomes the dominant species. *Acartia spp.* are the dominant genus in most neritic (shallow) water of the USA (Roman 1977). *Arcatia spp.* may be able to dominate these estuarine areas due to their wide salinity tolerance (Raymont and Miller 1962, Stein 1981). Further *A. tonsa* is a opportunistic, omnivorous, particle grazer and suspension feeder (Poulet 1978; Lonsdale et al. 1979). Omnivores have a distinct advantage to survival and dominance over more specialized feeders.

Size and Shape

The diversity of body shapes is also vast, from football-shaped to crab-like to worm-like (Figure 8.1). The body of most copepods is cylindriconical in shape, with a wider anterior part. The trunk consists of two distinct parts, the cephalothorax (the head is fused to the first six thoracic segments) and the abdomen, which is narrower that the cephalothorax. The head has a central nauplier eye and unirameous first antennae that a generally long.

There are over 4500 species in the Copepoda, ranging from microscopic to individuals as large as 50 mm (1/2 inch). Most adult copepods have a length between 1 and 5 mm. Males are lower in abundance and usually smaller than females. Sizes of copepods depends on the species as well as on the growth stage. A population of cultured *Euterpina acutifrons* (Calanoid) were measured in various stages of development and ranged from 50 x 50 x 75 μm for early instar stages to 150 x 150 x 700 μm for adults (Leighton, 1981). *Eutemora sp.* (Calanoid) are on the average 220 μm for nauplii, 490 μm for copepodites and 790 μm for adults. The Hepacticoid copepods *Tisbe holothuriae* grows from a nauplius size of 55 μm to an adult size of more that 180 μm, *Schizopere elateniss* from 50 to 500 μm and *Tisbentra elongata* from 150 to more than 750 μm.

Prey Value

Due to their overall larger size, copepods could be an important transitional food and/or a direct substitute for *Artemia*. Newly hatched copepod nauplii, range from 30 to 80 μm, and can be selectively sieved to feed small early stage larvae. There is little question about their food value since they are rich in waxy esters and marine oils. Food reserves are stored as oils in some copepods often giving the body a brilliant red color, as in *Diaptomus* (Barnes, 1963).

Stomach analysis of adult and subadult marine clownfish, *Amphiprion chrysopterus*, *A. melanopus*, and *A. tricinctus* revealed that copepods constitute 38% of their diet (Allen, 1972). There is no doubt about the food value of copepods in the culture of clownfish and many other marine fish however, enriched rotifers have been successfully substituted to rear many species of fish (Hoff 1996). Recent work on cod larvae by Hamre et al (2008) on enrichment of rotifers have resulted in values equal or better than copepods (see rotifer enrichment section).

Locomotion plays an important role in prey attraction. Like rotifers, copepods are slow swimmers. In addition, they move in sudden jerky motions which propel them forward up into the water column. Many calanoids display a slow gliding and jerking movement, resulting from the feeding current produced by the appendages around the mouth and propulsions caused by the thoracic limbs. Typically the orientation of the body when swimming is at 45°.

Culture copepods have been used successfully in Asia and Europe on several pelagic larval stages of various flatfish and turbot. Fukusho et al., (1980) fed mud dab larvae to 30 days on yeast fed cultures of *Tigriopus japonicus* with excellent survival and growth rates. Nellen et al., (1981) showed that after 14 days culture of turbot they shifted their feeding preference towards adult copepods over rotifers. Survival rate using copepods was 50% and the fry reached 12 mg dry weight and 17 mm total length in 26 days. Ogle (1979) successfully reared several thousand red snapper and suggested a 100:1 ratio of food to larvae during the larval rearing stage.

Feeding

Planktonic copepods are mainly suspension feeders on phytoplankton and/or bacteria. Food items are collected by the second maxillae. Copepods are considered selective filter-feeders. A water current is generated by the feeding appendages over the stationary second maxillae which actively captures the food particles (Lavens and Sorgeloos. 1996). When feeding, the second antennae sweep backwards and forwards very rapidly to generate a current of water which flows through combs or fine setae (hair or bristlelike structures) on the first and second maxilla. These setae remove potential food particles from the water and transfer the food to the mouth. Under optimal feeding conditions they produce fecal pellets that are coated with a chitin membrane at a rate of about 20 minute intervals (Marini 2003).

Species cultured

Most culture attempts have not been directed towards mass culture, but rather to provide a small amount of test animals for research (Zillioux & Lackie, 1970; Mullin & Brooks, 1967; Corkett, 1967). The calanoid copepod, *Acartia tonsa*, is more commonly cultured than any other species (Zillioux, 1969; Paffenhofer & Harris, 1979; Ogle, 1979; Ogle, 1999; Turk, et al. 1982). Leighton (1981) has conducted work on *Euterpina acutifrons* at the Waikiki Aquarium. In addition, Zurlini, et al. (1978) worked on the reproduction and growth of *Euterpina*. Kraul (1981) worked on the culture of *Oithona* and *Tisbe,* but found them unsuitable for aquaculture.

Perhaps the most success in culturing copepods as feed for larval culture has been with the harpacticoid *Tigriopus japonicus* (Hagiwara, et al. 1995b). Methods for its mass culture have been described by Fukusho (1980) and Kahan et al. (1981) and its physiological requirements by Lee and Hu (1980). Chandler (1986) described mass culture methods for other harpacticoid copepods and Stottrup and Norsker (1997) described a system for the harpacticoid *Tisbe holothuriae*. Success in the mass production of the calanoid *Acartia tsuensis* has been reported by Stottrup, et al. (1986) as well as Ohno and Okamura (1988). Lamm (1987) provided a simplified method to rear small numbers of copepods to a maximum density of 1/ml.

Kuhmann et al. (1981) successfully made harvests from 7.5 to 10% daily from a 24 cubic meter tank (6400 G) to feed tubot larvae. After 4-6 weeks of culture densities were estimated at several hundred adults and several thousand nauplii per liter. Although this was a success they were not able to stabilize the cultures and develop a reliable method. They estimated that this culture density was only sufficient to feed 4,000 tubot larvae until metamorphosis (24 to 30 days).

Lavens and Sorgeloos, (1996) reported maximum densities of *Acartia tsuensis* 1,300 nauplii, 590 copepodites with a maximum egg production of 350 eggs/L/day. They used a simplified synchronized method of placing nutrified, filtered (350-500μ) natural seawater in a tank then allowing it to go the various stage of development from diatoms to nanoflagellates and dinoflagellates followed by blooms of ciliates, rotifers and eventually copepods. Fish larvae were added when they determined suitable plankton levels were established.

McKinnon et al. (2003) investigated the suitability of three tropical species of small calaniod copepods for larval culture, *Bestiolina similis, Parvocalanus crassirostris*, and *Acartia sinjiensis*. All three species were easy to maintain in culture, but the nauplii of the first two species were smaller and easier for larval fish with small mouths like snapper and grouper to ingest. Algae *Rhodomonas* sp. and *Heterocapsa niei* produced the best copepod growth rates and acceptable DHA/EPA/ARA lipid ratios.

Life Cycle

Calanoids reproduce sexually. It appears that sex pheromones drive the primary male attraction in the mating process of calanoids (Chow-Fraser and Maly 1988). Males, upon finding a female, change their swimming motion from a smooth pattern to short bursts toward the female, catching her caudal rami with his geniculate right antennule. During copulation, males grasp the female with his fist antennae and deposits the spermatophores into seminal receptacle openings. Alignment between the sexes varies between species, but all use the right branch/ramus on leg 5 to hang onto the female's urosome and the left ramus of leg 5 to attach a freshly expelled spermatophore onto the genital segment near the genital aperture of the female. In some calanoids, there appears to be an optimal sex size ratio (female: male lengths) around 1.20, where successful copulation and clutch formation is maximized (Grad and Maly 1988). Mating may last brief seconds or require many hours.

Calanoid spermatophores are oblong or club-shaped, ending in a more or less elongated neck which is attached to the female's genital segment, and sealed with a glue plug. Females may have several spermatophores attached in the gonoporal area, but only the contents of one of these spermatophores is used to fertilize the eggs. For most species, it appears that a second mating is required for the females to produce eggs again (Dussart and Defaye 2001).

Females of most calanoid species carry one egg sac (ovisac) or brood chamber, containing from 3-35 eggs, located in the middle of the first abdominal segment (urosome area). Eggs are usually retained there until hatch. Some species scatter their fertilized eggs freely into the water, where they hatch later. The development time of diaptomid subcutaneous eggs is primarily a function of temperature and can vary from 2-35 days (Herzig et al. 1980).

In contrast to rotifers that reproduce mainly asexually and daphnia, copepods reproduce sexually. Rotifers begin reproduction about 18 hours after hatch however, copepods, can take a month or more to complete their reproductive cycle. Because of this, it is difficult to maintain a culture that can produce sufficient offspring to sustain a large population of rapidly growing nauplii (Moe, 1989). *Acartia tonsa* survive up to 24 days as adults and produced up to 75 nauplii per female adult (Ogle, 1979).

Female copepods are capable of retaining unfertilized eggs for extended periods. Fertilized, eggs (oocytes) are secreted through the oviducts into sacs (ovisacs) which serve as a brood chambers for a few to forty eggs. Brood chambers are attached to the females first abdominal segment. Multiple ovisacs can be laid after a single fertilization. Only a few species release the fertilized eggs directly into the water column. Most species retain egg sacs externally under the body until hatch. Hatching of the freshwater *Cyclops*, takes from 12 hours to five days. Male copepods are commonly smaller and less abundant than the females. During copulation the male grasps the female with his first antennae and deposits the spermatophores into a seminal receptacle opening where they are glued by means of a special cement. Calanoids shed their eggs singly into the water. Fertilized eggs hatch into nauplii which undergo five to six molts called naupliar stages and then become copepodites. After another five copepodite molts they become adults and molting ceases. Development may take less than one week to a year depending on specie and environmental conditions. Life span may range from six months to one year. Under ideal conditions and temperatures of 24°-26°C (75°-79°F), *E. actifrons* undergoes a series of growth molts which includes 6 naupliar stages and 6 copepodite stages (including adult). Full development takes about 8-11 days to maturity (Lavens and Sorgeloos.1996).

Marina (2003) described egg development in more detail. Newly deposited eggs are dark brown and as the embryos mature the colors change to a light brown and have a visible dark eye spot. Nauplii emerge from the egg sac and contain four or five oil droplets which serve as an initial energy source. Development time is temperature dependent. At 25°C (77°F), embryo and nauplius stages are cumulated in 4-5 days and full maturity (embryo to adult) takes a total of 10-12 days.

Resting Cysts

Like brine shrimp and rotifers, copepods undergo a period of suspended development (diapause) in the form of resting eggs as a common life-cycle strategy to survive adverse, unfavorable or less than optimal environmental conditions. Under these conditions both freshwater and saltwater copepod species can produce thick-shelled dormant eggs (resting eggs or cyst) which sink to the bottom awaiting optimal conditions for hatch. These special eggs can withstand desiccation and also provide a means of dispersal via birds and animals. Experiments have shown that resting eggs can tolerate drying at 25°C (77°F) or freezing down to -25°C (°F) and can resist these temperatures as long as 9 to 15 months. Producing resting stage cysts (diapause and non-diapause) is more common in upper northern regions in order to survive adverse environmental conditions, such as freezing. Diapause usually takes place between the copepodite stage II to adult females which are recognized by an empty alimentary tract, the presence of numerous oil globules in the tissue and a organic, cyst-like covering. Most diapause cysts are in the sediment although some can remain within the water column as planktonic fraction and are referred to as "active diapause" (Lavens and Sorgeloos.1996). Some data indicate that eggs can survive for many years in sediments which would expand their influence to evolutionary time scales (Marcus and F Boero 1998).

Viability of long term storage of resting eggs is tenuous and variable. Hojgaard et al (2008) tested changes of percent hatch and storage time of *Acartia tonsa* cysts under temperatures of 17°C and 25°C and salinities from 0 to 30 ppt. (parts/thousand) Dormancy was strongly influenced at 0 salinity and partially at 5 salinity in both temperatures. Eggs stored at 0 salinity and 17°C remained 25% viable for 12 days whereas those stored at 25°C which showed a gradual decline in viability until stabilizing at 10% from day 7 onwards.

Food Requirements

Nutritional Quality

Like rotifers copepods are what they eat. Copepods have a high protein content (44-52%) and a good amino acid profile with the exception of methionine and histidine. Fatty acid composition varies considerably which reflects composition of their diet. For example, the (ω-3)HUFA content of individual adult *Tisbe* fed on *Dunaliella* which has a low (ω-3)HUFA content and *Rhodomonas* which has a high (ω-3)HUFA is 39 ωg and 63 ωg respectively which corresponds to 0.8% and 1.3% of dry weight. Nauplii levels were relatively higher ranging around 3.9% and 3.4% respectively. Specific levels of EPA and DHA are respectively 6% and 17% in adults fed *Dunaliella* and 18% and 32% in adults fed *Rhodomonas*. Nauplii levels were around 3.5% and 9.0% when respectively fed both algae (Lavens and Sorgeloos.1996).

Compared to Artemia the HUFA content of copepods is higher. In addition both the copepodite and adult stages of copepods are believed to contain higher levels of digestive enzymes which may play an important role during larval nutrition (Lavens and Sorgeloos.1996).

Khanaichenko (1998) found that mixed diets including dinoflagellates enhanced reproduction and survival patterns of copepods increasing their DHA up to 14.0-16% on a monodiet to 25.8% on a mixed diet. DHA/EPA increased up to 3.26-3.68 on a mono diet to 6.14 on mixed diets. The mixed diets were not clearly specified nor was the dinoflagellate specie.

Microalgae

Herbivorous copepods are chiefly filter-feeders, but some species are predators that can capture and consume larval fish. Planktonic copepods are the most suitable for larval culture. In contrast to rotifers, diatoms rather than green algae constitute the principal part of the copepod diet. Kraul (1981) preferred *Chaetoceros* sp. for copepod culture and suggested that feed levels should not exceed 200,000 cells/ml for *Euterpina acutifrons*. Levels above 500,000 cells/ml seemed to suppress growth. He further suggested that the addition of the dinoflagellate *Gymnodinium* increased fecundity and produced faster growth. Optimum concentrations of food organisms were selected using results from carbon-14 labeled phytoplankton on *Acartia clausi* adults (Zillioux, 1969). When fed *Rhodomonas baltica*, maximum efficiency was achieved when the food concentration was 50,000 cells/ml. When fed *Isochrysis galbana,* the ingestion rate was considerably slower until the cells began to clump together at levels well in excess of 50,000 cells/ml. Since the test animals were adults, he speculated that the smaller *Isochrysis* cells were more suitable for the earlier naupliiar stages. From this research it was suggested that both algae species should be provided and not exceed a concentration of 100,000 cells per ml.

Apeitos et. al. (2004) conducted microalgae feeding studies on *Acartia tonsa* using a golden -brown algae (*Isocrysis galbana*) and a diatom (*Chaetoceros sp.*). Mature female copepods (100) were placed individually in 10 ml containers. Twenty five were fed *Isocrysis* (150,000 cells/ml), twenty five fed Chaetoceros (150,000 cells/ml), twenty five fed a combination diet of both algae (75,000 cells/ml of each), and twenty five were not fed. Eggs were routinely collected and counted over 72 hours. Two sets of experiments were conducted. Results indicated that the survival (68%) and fecundity was higher in the mixed diet where 17 of the 25 produced a total of 520 in 72 hours.

Cotonnec et al (2001) identified chemical biomarkers (pigments and fatty acids) of microalgae consumed by the dominate copepod species *Temora longlcornis, Arcartia clausi* and *Pseudocalanus elongatus* in the English Channel. Studies were conducted over a three day period with samples collected every three hours. It was performed during the spring phytoplankton bloom when the microalgae *Phaeocystis sp.* formed 90% of the total phytoplankton. Their study showed that these specific zooplankton species selectively grazed on *Phaeocystis sp.*, non-selectively on diatoms while dinoflagellates were avoided. *Temora longlcornis* selectively grazed on Cryptophytes (colony forming diatoms) which is probably related to their high nutritional value. Fatty acid composition of the three copepod species indicate an herbivorous diet for *P. elongatus* and an omnivorous one for *A. clausi* and *T. longicornis* which is the least opportunist.

Ogle (1979) utilized natural filtered (5 μm) seawater from the Mississippi Sound which is brown colored due to rich blooms of diatoms and golden brown alga. Chlorophyll a determinations over several years, were comparable or exceed values for algae diets used to culture oyster larvae.

Fengqi (1996) made notation that in the pond culture of rotifers in China, copepods are predators. Therefore, it is possible that rotifers should be considered as an alternative food source when culturing copepods.

Work in Australia (1999) on the culture of calanoid copepod (unidentified) in 500L tanks showed that a diet of *Isocrysis* promoted high nauplius production but warned that it should not be used alone for long periods of time. When a dinoflagellate (*Heterocapsa niei*) was added to the diet menu copepod maturation time was faster using this diet alone. Indicating it may have the potential to improve productivity.

Khanaichenko (1998) also recognized dinoflagellates as a monodiet or used in mix diets. Mixed diets allowed for maximum production of eggs. In this work they collected the unhatched eggs every 12 hours and refrigerated them to 4°C. Collecting and cold storage of eggs would provide them with a larger amount of prey size nauplii (110-260μm) when needed. They simply placed them in 20-24°C water and would hatch in 17-26 hours at a rate of 55-92%. This a very unique approach to providing numbers of right size copepods when needed.

Brans, Meals, etc.

More recent work by Ogle et al. (unpublished) indicated that *Acatia tonsa* is more inclined to be a opportunistic suspended particle feeder. In laboratory studies, feeding detritus had not been generally successful whereas, use of phytoplankton is standard. Energy studies on feeding in a number of copepod species suggested that in the wild they can not derive enough energy from grazing exclusively on phytoplankton (Heinle and Flemer, 1975). It has been demonstrated that copepod adults and nauplii can utilize detritus and indeed there is 5 times more detrital than phytoplankton biomass in the ocean (Poulet,1977). It has been suggested (Poulet,1983) that free bacteria are not used as a food by copepods and there is 200 times less bacteria biomass in the ocean than detrital biomass. Takano (1971) reared the estuarine calanoid *Gladioferens imparipies* on a wheat and soya flour diet. He speculated that bacteria growing on the food particles rather than the particles themselves, was the main food source for the copepods. Turk, et al. (1982) utilized defatted rice bran sieved through a 73 μm mesh to rear *Acartia tonsa*. It was mixed in deionized water at a concentration of 1 to 3 mg of rice bran per liter of culture water and was offered twice daily. Feeding levels were reduced if the culture water did not clear between feedings.

Particle feed experiments were conducted on *Acartia tonsa* by . Tanks were 1000L (250 gal.), 12 total with 3 replicates for each treatment, 3 trials, fed every third day. One trial consisted of feeding algae paste of *Tetraselmis sp.*, a bacteria floc used in shrimp water, Isomil™ and shrimp culture water and a unfed control. Two trials consisted of three diets, Roti-Rich™ (Florida Aqua Farms), rice bran and Artificial Plankton™ (Aqua-Fauna) and a unfed control.

Results showed that the number of nauplii increased when fed, shrimp waste water, bacterial floc, Artificial Plankton™, Roti-Rich™, and rice bran. The number of nauplii continued to increase in tanks being fed shrimp water while a decline during the fourth week was noted for tanks being fed Artificial Plankton™, Roti-Rich™ and rice bran. Greatest increase in adults were noted for the artificial plankton (2,306%) during the fourth week and Roti-Rich™ (2,194%) during the third week representing 240 and 121.6 adults/L respectively. Overall greatest increase in the number of nauplii was noted for the Roti-Rich™ during the third week (937%). This percent represents 120.3 nauplii/L which was the second highest density. The greatest density of nauplii achieved was 171/L also for Roti-Rich™ in the third week. However, due to the high initial stocking density in this trial (118 nauplii/L) this represents only a 44.9%. Roti-Rich™ fed tanks had the greatest percent increase in nauplii, greatest total number of nauplii/L and second highest percent increase in adults shows promise as an artificial feed. The microalgae fed was preserved and not living and preformed poorly. However, the algal cells were considered as a detrital particle and some success was expected. The poor performance of bacteria was not surprising as bacteria are not considered a major food of copepods (Poulet 1983). These experiments clearly demonstrate that *A. tonsa* are omnivorous particle feeder. Results of this study confirms that *Acartia tonsa* is indeed an omnivorous feeder able to feed and reproduce on several diets offered.

Lemus et al. (in revision) conducted feeding studies utilizing "brown water" from the Mississippi Sound and rice bran. Brown water is named for its color, which results from suspended phytoplankton, detritus, dissolved organics and other non biological sediments. It contains chlorophyll-a and nitrate levels similar to cultured phytoplankton water. While the primary productivity of the natural estuary is high, it remains to be seen whether supplemental feeding could improve copepod production. This water was strained through a 5 μm mesh bag. Twelve 1000 L circular tanks were use for the first experiment and twelve 200 L for the second. In each experiment three feeds were tested and one unfed control. Rice bran (5 gm/day) was provided to 3 tanks; rice bran (5 gm/day) and 50% brown water exchanges were made every other day; 3 tanks just had brown-water exchanges every other day and the remaining 3 tanks were controls that received no food or brown-water.

Brown-water cultures that were fed rice bran had significantly greater copepod nauplier densities than cultures fed only rice bran, only brown-water and the controls. The goal of the second set of experiments was to determine the harvest rate (25%, 50% and 75% every other day) that would supply the most copepod nauplii. Naupliar yields increased with increasing harvest levels. There were significantly lower yields at a 25% harvest level, but no significant difference in naupliar yields between a 50% and 75% harvest.

Physical Requirements

Light

Zillioux (1969) used constant overhead illumination of approximately 1500 lux from a single 40 watt cool-white fluorescent bulb to rear *Acartia*. He felt constant light keeps the phototactic (attracted to light) populations generally well suspended in the water column. Turk, et al. (1982) also used constant low light provided by a single 40 watt fluorescent bulb suspended 20 cm above the tank. Kraul (1981) suggested partial shading of culture tanks helps when rearing *Euterpina*. Based on these reports and our experience with rotifers, we suggest using constant, low level light.

It is possible that low or no light may be beneficial. We have found copepods actually surviving and possibly multiplying inside cartridge filters. The housings are opaque light tan PVC pipe that allow low levels of light to pass through. Since many natural populations of copepods have diurnal migration patterns it is possible that low light might be more beneficial.

Ogle et al. (1999) conducted experiments in greenhouses with 85% shade cloth.

Aeration

Turk, et al. (1982) placed two, 25 mm (1") diameter air lift tubes with 90° elbows on top in opposing corners (see Figure 7.4). No mention was made about the amount of air supplied but, based on the size of the culture tank, 100 x 50 x 50 cm (40" x 20" x 20"H) the circulation and flow could be substantial. Too vigorous aeration damages fragile copepods and may cause premature release of eggs (Lamm, 1987). Ogle (1979) on the other hand, stated that the fragile sensitive qualities noted by other authors was not evident since he routinely sieved and concentrated his cultures with little loss. We recommend moderate aeration perhaps without an airstone. More than one outlet may be necessary in larger tanks to provide an uplift circulating effect.

Culture tank size

There does not seem to be an optimal tank size or configuration, but several points should be considered. Depth is probably not as important as surface area. Higher surface promotes better gas exchange, thus higher oxygen levels without extreme aeration. Ogle (1979) used 1800 L (500 gal.) circular tanks with good success but more recent work provided good results in 1000L (250 gal .) tanks. Whereas, Lamm (1987) used plastic garbage cans and Turk, et al. (1982) used rectangular tanks as noted previously.

Filtration

As compared to rotifers and daphnia, copepods are not as tolerant of high particulate matter, dissolved organics, or high bacteria levels. Lamm (1987) utilized a small air-lift sponge filter in his cultures. Zillioux, (1969) utilized a protein skimmer to reduce dissolved organics and ciliates to reduce bacteria levels in a closed culture system. Ogle (1979) made complete water exchanges three times weekly, drained the tanks, selectively collected various sizes of copepods, cleaned and dried the tanks, and re-stocked again. Based on this evidence, it is suggested that reasonably clear, well aerated water is desirable. Use of protein skimmers to reduce dissolved organics in closed systems is strongly suggested, but use of sponge filters in culture tanks can be problematic, especially when feeding microalgae and micronized foods. Use of sponge filters might be possible if they are used only during non-feeding periods.

Temperature

Kasahara et al, (1975) working with resting eggs of the copepod *Tortanus forcipatus* found that hatching occurred at 13° to 30°C and no eggs hatched at 10°C. Optimal temperatures were around 25°C although the older the resting eggs the lower the optimal temperature for hatch. Kraul (1981) suggested temperatures of 25°C ± 3°C for rearing *Euterpina acutifrons*. Zillioux (1969) maintained cultures of *Acartia clausi* and *tonsa* at 15°C, but did not explain why he used such low temperatures. He felt that excessive bacteria were detrimental and the lower temperatures would no doubt slow bacteria growth. Ogle (1979) cultured *Acartia tonsa* over a 6 month period during which water temperatures ranged from a low of 5.5°C to a high of 27.7°C with an average of 20°C. Ogle (2000) reared *A. tonsa* from January to April at 6°C to 16°C. We suggest a range from 12° to 26°C.

Salinity

During a 6 month culture period, natural salinities ranged from 1 to 26 ppt, (1.0000-1.0145 density) with an average of 12 ppt (1.0084) (Ogle 1979). Since large concentrations of copepods are found seasonally in bays and estuaries, culture salinities should be maintained in mid-range, from 12 to 20 ppt. (1.0084 to 1.0176 density). Maintaining cultures at higher salinities 30-35 ppt (1.0222 to 1.0260 density) could promote contamination of less desirable harpacticoid species. However if you are working with a harpacticoid species a higher salinity is desirable.

Oxygen

As compared to rotifers, daphnia and *Artemia,* copepods cannot tolerate low levels of oxygen. Zillioux (1969) lost cultures of *A. clausi* when the oxygen levels dropped to 3.2 ppm. After removing the dead copepods and increasing the aeration, stage 1 and 2 nauplii were noted one day later which indicated that certain stages can tolerate lower oxygen levels. High levels of oxygen are recommended.

pH

Compared to rotifers and daphnia, copepods are not tolerant of high dissolved organics which reduce pH levels. Therefore, pH should be maintained at normal seawater levels of 8.0 to 8.2.

Culture Methods

Culture Equipment

Many species of copepods are found in the upper water column or in shallow bays. The specie being cultured may dictate what type of tank may be required. However, a shallow water column (less then 36") and a wide surface area are recommended. Color of the sides of the tank may be a consideration. For a rectangular tank (aquarium) you may want to cover three sides with black plastic and white styrofoam under the bottom. Light should be provided above the aquaria. For small cultures a simple setup like that shown in Figure 5.5 (Rotifer section) should suffice except a separate light above the culture may be required.

Culture Parameters

Culture methods differ, but in general are not complex. Yet, methods cannot be as lax as growing rotifers. The following procedure is only a guideline.
1. Tank size is variable but a high surface area like an aquarium is preferred.
2. Provide moderate aeration without an airstone in one corner of the tank to provide circulation.
3. Copepods prefer higher oxygen levels 3.2+
4. Preferred pH range 8.0 to 8.2 for saltwater cultures.

5. If a saltwater culture keep salinity below 30 ppt. Preferred range 12-20 ppt. (1.0084-1.0145).

6. Temperature range can vary from 12° to 27°C (54°-81°F).

7. Use constant, low level light above the culture.

8. If feeding microalgae, diatoms or dinoflagellates are apparently better than green algae.

9. Utilizing a combination of live microalgae and non-living micronized foods will no doubt produce the highest yields.

10. Non-living fine ground larval diets consisting of, meals or brans, or yeast-based foods like liquid and dry Roti-Rich™ can be used. Care must be taken to avoid overfeeding which results in high bacteria levels.

11. Feeding should be minimal at first and increased as the population increases.

12. Routine water exchanges and minimal bottom siphoning are strongly suggested.

13. Culture tanks and water must be kept relatively clean.

14. Short-term, continuous cultures are probably the best way to culture copepods.

15. Tanks, as with rotifer cultures, should be dried and cleaned every 2 to 4 weeks and restarted with 10% old and 90% new water plus microalgae an a inoculant of concentrated adult animals.

16. Rules utilized in rotifer culture should apply, except higher water quality standards should be maintained.

Harvesting and density

Harvesting will depend on what size range you need. Ogle (1979) harvested various stages of co-pepods utilizing sieves of various sizes. A 200 - 250 μm sieve will retain only adults, allowing nauplii and copepodites to pass. The copepodites can be retained by a 100 μm sieve and the nauplii by a 25-45 μm sieve. Harvesting can occur daily, but care should be taken not to remove too many of one size class. Random collections and/or collections of only sub-adults are probably the best way to maintain the highest densities. Average production in Ogle's system was 232,000 adults per m^3 which is 580 times higher than the adjacent bay from where the culture water was removed.

Turk, et al. (1982) report maximum densities of *Acartia tonsa* in four experiments ranging from 870 to 1,680 nauplii per liter (avg. 1,225/L) and 170 to 1,520 copepodites, with adults averaging 679 per liter. Time required to reach maximum density ranged from 6 to 33 days for nauplii and 12 to 27 days for copepodite stages.

Obtaining starter cultures

If you can obtain fresh "live-rock," chances are good you will obtain some desirable copepods. Another possibility is mud or bottom samples from a healthy bay. Sieve the mud and clean as much as possible. Resting eggs may be present. A third possibility is obtaining concentrated open ocean or, preferably, bay or estuary water from the coast. Place copepods in clean water, aerate, feed micro-algae, and wait. *Acartia* are often seen concentrated in the corners of a well balanced aquaria during certain periods of the day. They often rest on the glass. Siphon them out, place in a tank, and feed microalgae.

Another source is obtaining resting eggs from bottom sediments. Untreated sediments can be stored at 2-4° C (36-39°F) for several months prior to use. Naess and Bergh (1994) discuss other methods of processing and treating copepod resting eggs. When needed sediment is put into liquid suspension and sieved through 150μ, 100μ and 40μ sieves. That which is retained on the 40μ screen Eggs can be removed by reconstituting the 40μ mass in clean sterile saltwater and place in shallow round bottom dish. This is then swirled, like panning for gold, until the eggs are concentrated and exposed on the bottom. These can be siphoned off with an eyedropper, placed in a capped vial with sterile saltwater then stored under dark, dry cool conditions. Work on the controlled production of resting eggs from the calanoid *Centropages hamatus* is progressing (Marcus and Murray 2001) and these may become available for use as starter cultures in the near future. Live inoculant cultures can also be obtained from several companies including Florida Aqua Farms, Carolina Biological Supply, Wards Scientific etc.

Mosquito Control

Although parasitic copepods are not desirable for use in larval rearing they may be used in other ways including contaminate control or usable in juvenile rearing. Biological control is a more attractive alternative then chemical pesticides. One recent alternative is the culture of the predatory freshwater cyclopoid copepod *Macrocyclops albidus*. Preliminary results indicate that this predator is highly sufficient in controlling mosquitoes. In field studies they have demonstrated that there was a 90% reduction in mosquitoes. This predator copepod is most efficient on 1-4 day old larvae.

M. albidus is cold tolerant and once introduced the population can survive for moths at 0°C. The specie has a worldwide distribution so seed stocks should be easy to obtain, and practical and legal problems associated with exotic species would not apply. Large numbers of copepods can be kept in water in a refrigerator for months, they can survive in soil and detritus that is only slightly damp. In addition they are not killed by many pesticides commonly used for mosquito control.

One method of culture is uses *Paremecium* as a food source (see *Paremecium* culture). In this example the *Paremecium* are reared in conjunction with the copepods. Wheat seed, a food source for the *Paremecium*, are fed into a 80L pool which has also been stocked with copepods. As the wheat seed decomposes, they float to the surface and are removed. These culture pools are emptied and cleaned at least once a year. Adults, copepodites, nauplii

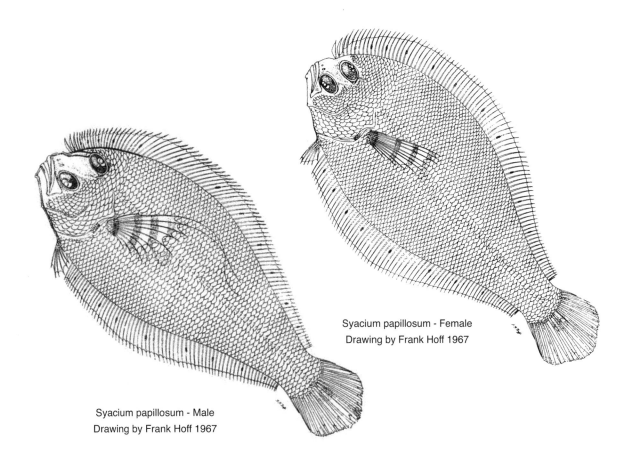

Syacium papillosum - Female
Drawing by Frank Hoff 1967

Syacium papillosum - Male
Drawing by Frank Hoff 1967

Chapter 9 - DAPHNIA CULTURE

Introduction

Daphnia are small freshwater crustaceans commonly called "water fleas." This common name is the result not only of their size, but their short, jerky, hopping movement in water. Many species of the freshwater family Daphniidae (order Cladocera) occur throughout the world and are collectively known as daphnia. The genera *Daphnia* and *Moina* are closely related cladocerans but are collectively known as daphnia. *Daphnia* and *Moina* are the most diverse of the daphnids and are a major food for both young and adult freshwater fish. Literature abounds on the capture, biology, culture, and use of *Daphnia* sp., but little exists on *Moina*. Yet, within the Oriental countries, *Moina* is the number one live food for rearing freshwater tropical fish. In Singapore, *Moina micrura* are grown in ponds fertilized with chicken or pig manure (May, et al. 1984) and are used as the sole food for post-larvae of many ornamental fish, with 95 to 99% survival to 20 mm in length quite common. *Moina* also has been used as an *Artemia* substitute in the production of post-larvae of the freshwater prawn *Macrobrachium* (Alam, et al. 1993a & b). In addition, daphnia are used to inoculate production ponds for sport fish such as striped bass (*Morone saxitillis*). Pond production in Singapore utilizing re-cycled hog wastes produces 24 Kg (528 lbs.) of *Moina* per 1500 M³ (400,000 gal) at each harvest (Lee Heng Lye, 1983).

Growth experiments showed that newly hatched freshwater angelfish grew 32% longer on *Moina* compared to newly hatched *Artemia* over a 15 day period. At 30 days, length was about equal, but body depth was larger on *Moina*-fed fish. Color was significantly better developed on 30 day old, *Moina*-fed fish (Hoff and Wilcox ,1986.).

Natural populations

Natural populations of *Daphnia* and *Moina* occur in high concentrations in pools, ponds, lakes, ditches, slow-moving streams, and swamps where there is decomposing organic material. They become especially abundant in temporary waters which provide them with optimal growth conditions for brief periods. Different genera and species of daphnia, however, have different environmental requirements. *Moina sp*. are generally more tolerant of poor water quality than *Daphnia sp*. They live in water where the amount of dissolved oxygen varies from almost zero to supersaturation.   *Daphnia magna* and *Moina macrocopa* are particularly resistant to changes in the oxygen concentration and often reproduce in large quantities in waters polluted with sewage. Species of *Moina* have been reported to play an important role in the stabilization of sewage in oxidation lagoons (Loedolff, 1966).

Size and morphology

Considerable size variation exists between and within the genera. Adult *Moina* (700-1500 μm long) are larger than newly hatched brine shrimp (400-500 μm), and approximately 3 to 5 times the length of adult rotifers. Newly hatched *Moina* average 430 μm and are slightly larger than the smallest strain of newly hatched brine shrimp, and twice as large as adult rotifers. Adult female *Moina* are larger and rounder than males, with an average size of 1.25 mm, (700-1250μm) while males are about 0.6 mm (600μm). *Moina* are about half the length of *Daphnia pulex*. Young *Daphnia* as a result of their large size, are not suitable for the fry of many fish larvae. However, newly hatched larvae or mid-stage larvae of many species can ingest young *Moina* as their initial food.

Daphnia have a body consisting of a trunk and a head with five appendages. The first pair, the antennules, are rod-like and show sexual dimorphism. The second pair are antennae and the main means of locomotion. The last three pairs of appendages are used in feeding. Large compound eyes lie under the skin on either side of the head (Figure 9.1). One of the major characteristics of daphnia is that the trunk is enclosed in an external skeleton (carapace). Five pairs of legs are enclosed by the carapace. The brood pouch, where the eggs and embryos develop, is on the dorsal side of the female. In *Daphnia*, the brood pouch is completely closed, while *Moina* have an open pouch (Rottman 1992). As they grow *Daphnia* molt or shed their external shell (Ivleva, 1969).

Marine Cladocerans

Cladocerans mainly inhabit freshwater, but a few species are marine. Some have been investigated for their suitability as mass cultured live foods in aquaculture. An example is *Diaphanosoma aspinosum* which has been cultured at reasonably high densities by Segawa and Yang (1988). They report achieving densities of 131/ml after 32-42 days at 25°C and 15 ppt on a diet of the green alga *Tetraselmis chui*. Culture water was exchanged daily. Neonates are about 0.5 mm long, so they are approximately the same size as newly hatched *Artemia* nauplii. Adults typically grow to 1.3 mm in length (Segawa & Yang, 1990). Other species with promise are marine *Moina salina* (Gordo, et al. 1994) and perhaps *M. micrura* in brackish water (Bonou and Saint Jean 1998), but more work is needed on all three species before their mass culture is productive and reliable enough for use in commercial aquaculture.

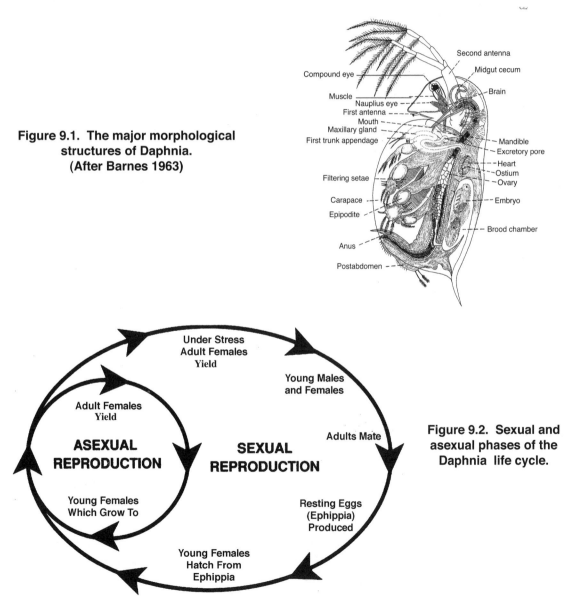

Figure 9.1. The major morphological structures of Daphnia. (After Barnes 1963)

Figure 9.2. Sexual and asexual phases of the Daphnia life cycle.

Life Cycle of Daphnia

The reproductive cycle of *Daphnia* and *Moina* have both sexual and asexual phases. In most environments, populations consist exclusively of females that reproduce asexually (Figure 9.2). Under optimum conditions, it has been reported that *Daphnia* can produce more than 100 eggs per brood and a brood every 2.5 to 3 days. A female can have as many as 25 broods in her lifetime (Ivleva, 1969). *Moina* reproduce at 4 to 7 days of age, with a brood size of 4 to 22 per female. Broods are produced every 1.5 to 2.0 days, with most females producing 2 to 6 broods during their lifetime (Ivleva, 1969; Conklin & Provasoli, 1977).

Resting eggs

Under adverse environmental conditions, males are produced and sexual reproduction follows, resulting in resting eggs (ephippia). The stimuli for the switch from asexual to sexual reproduction in populations of *Daphnia magna* include factors like food deficiency, oxygen starvation, high population density and low levels of calcium and perhaps other trace metals and minerals. In addition, low temperature and either extremely long (20 hr) or short (4 hr) photoperiod can trigger the production of ephippia (Ivleva, 1969). For *Moina*, an abrupt decrease in ingestion rate caused by a rapid reduction in the food supply results in increased ephippia production. Females eject fully developed ephippia during their molts. A female only has to be fertilized once which suffices for the next two or three broods of resting eggs.

Ephippia are black or dark colored, rectangular shaped, with rounded corners and about 1-2mm long. They can be found on the bottoms of ponds or along the shore. They can be air-dried and stored at 1-5°C (34-41°F). Ephippia of *Daphnia* remain viable even when frozen, but some species of *Moina* apparently do not survive freezing. To hatch ephippia of *Daphnia*, they should be placed in aerated water at 18-22°C (64-72°F). Hatching normally occurs in 4-7 days, resulting in females which begin asexual reproduction. Ephippia of *Moina* hatch in 2-4 days at 20-29°C (Ivleva, 1969). For most aquaculture applications, it is desirable to keep the population well fed to enhance asexual reproduction, since fewer progeny are produced with sexual reproduction.

Population densities

High population densities of *Daphnia* can result in a dramatic decrease in reproduction, but this is apparently not the case with *Moina*. The egg output of *Daphnia magna* drops sharply at population densities as low as 25-30/L (95-115/G) (Ivleva, 1969). The maximum sustained density in cultures of *Daphnia* was reported by Heisig (1977) as 500/L (1900/G). *Moina* cultures, however, routinely reach densities of 5000/L (19,000/G) and are, therefore, better adapted for intensive culture. A comparison of the production of *Daphnia magna* and *Moina macrocopa* cultures fertilized with yeast and ammonium nitrate showed that the average daily yield of *Moina* was 106-110 g/m^3 (1.03-1.07 oz/100 G), which is 3-4 times the daily production of *Daphnia* (25-40 g/m$^{3)}$ (Ivleva, 1969).

Physical and Chemical Requirements

Salinity

Daphnia are typically freshwater organisms, however, some are found in slightly brackish water. *Daphnia longispina*, *Moina rectirostris,* and *Moina macrocopa* have been observed in salinities up to 4 ppt (Ivleva, 1969). Salinities of 1.5 to 3.3 ppt are common in pond cultures in the Orient (Lee Heng Lye, 1983). The estuarine cladoceran *Diaphanosoma aspinosum* and *D. celebensis* can be mass cultured at salinities ranging from 1-42 ppt (Segawa & Yang, 1987; Achuthankutty et al. 2000). The marine species *Moina salina* also can be cultured at 36 ppt (Gordo, et al. 1994).

Oxygen

Daphnia are generally quite tolerant of poor water quality, and dissolved oxygen varies from almost zero to supersaturation. Their ability to survive in oxygen-poor environments is due to their capacity to synthesize hemoglobin. The rate of hemoglobin formation is dependent on a reduced level of dissolved oxygen in the water. The production of hemoglobin may also be promoted by high temperature and high population density. Kobayashi (1981) showed that the color of *Moina* is directly related to the hemoglobin (Hb) concentration which, like *Artemia*, is due to the oxygen content of the water and diet. Pink animals had 1.7 ± 0.2 g Hb/100 ml blood compared to 1/15 that in pale animals. The lower the oxygen level the more intense the color. Pink animals have excellent ability in removing oxygen from poorly aerated waters. Utilization of tanks with high surface area and shallow depths are recommended to facilitate oxygen exchange.

pH and ammonia

A pH between 6.5 and 9.5 is acceptable. High ammonia levels in conjunction with a pH of 8.5 or higher drastically reduce reproduction. Yet, *Moina* are tolerant of high ammonia levels. Information is somewhat conflicting, but it appears that levels of un-ionized ammonia of 15 to 20 ppm is acceptable for *Daphnia*. *Moina* tolerate higher levels ranging from 35-50 ppm (Lee Heng, 1983). These high levels of un-ionized ammonia help reduce contaminants. Other cladocerans are also tolerant of un-ionized ammonia. The 24-hr LC_{100} (concentration causing 100% mortality in 24 hr) of freshwater *Diaphanosoma* has been reported to be just under 20 ppm (Lincoln, et al. 1983).

Dissolved minerals and metals

In contrast to their tolerance of low oxygen, *Daphnia* are highly sensitive to disturbances of the ionic composition of the culture medium, especially changes in the concentration of some cations. They become immobile and die as a result of the addition of salts like sodium, potassium, magnesium, and calcium. Concentrations of only 0.01 ppm copper, for example, result in reduced movement of daphnia, whereas sodium ions are much less toxic (Ivleva, 1969).

Low concentrations of phosphorus (less than 0.5 ppm) stimulate reproduction of *Daphnia pulex*, while a concentration of 2 ppm is lethal to the young (Heisig, 1977). Best results are obtained if the phosphorus concentration is 1 ppm or less. *Daphnia magna* are more resistant to phosphorus. In intensive cultures, this species usually withstands a concentration as high as 5-7 ppm (Ivleva, 1969). *Daphnia* do not usually react adversely to an increase in the concentration of nitrogen as a result of fertilization.

The quality of the water source is important in *Daphnia* cultures. Well water should be aerated for at least two hours. Municipal water should be aerated for at least two days to reduce the chlorine concentration to non-toxic levels. Alternatively, sodium thiosulfate or a commercially available chlorine neutralizer can be added to shorten this process.

Daphnia are extremely sensitive to metal ions like copper and zinc, pesticides, detergents, bleaches, and other dissolved toxins, which may be present in shallow well and surface waters. Even municipal waters may present a problem and alternative sources may have to be used such as spring water, filtered lake or stream water, aerated well water, or rain water collected from areas with low air pollution. For aquarists, the easiest source of water may be from an established, disease and parasite free aquarium. If these are unavailable, try unaltered spring water, but not distilled water.

Temperature

Daphnia magna have a wide thermal range, withstanding daily changes from 0-22°C (32-72°F). The optimum temperature of *Daphnia magna* is 18-22°C (64-72°F) (Ivleva, 1969). *Moina macrocopa* are even more resistant to extremes in temperature and easily withstand a daily variation of from 5-31°C (41-88°F). Their optimum temperature is 24-31°C (75-88°F) (Ivleva, 1969). The higher temperature tolerance of *Moina* is of great advantage for aquarists culturing live food at home and commercial tropical fish farmers who maintain their hatcheries above 18-22°C (64-72°F).

Food Requirements

Daphnia feed on various groups of bacteria, yeast, microalgae (both single-cell algae and colonies of blue-green algae consisting of two or three cells), detritus, and dissolved organic matter (Monakov, 1972). An extended shortage of microalgae has been shown to adversely influence populations by reducing the number of young individuals (Heisig, 1977). Bacterial and fungal cells rank high in food value, but not as high as microalgae. Populations of *Daphnia* grow rapidly in the presence of adequate amounts of bacterial or yeast cells, as well as microalgae cells (Dewey and Parker, 1964). Both plant and animal detritus probably support the growth and reproduction of daphnia. The food value of detritus depends on its origin and diminishes with age (Pavlyutin, 1976).

Moina is one of the few zooplankters that can utilize the blue-green alga *Microcystis aeruginosa* (Hanazato & Yasuno 1984). Although the growth and reproduction of *Moina* was significantly lower when fed *Microcystis* alone, a mixed culture of *Chlorella* and *Microcystis* produced no significant inhibitory effect. Rao and Krishnamoorthi (1977) studied the preferential feeding habits of *Moina* in a mixed algae population in a waste stabilization pond having *Chlorella, Scenedesmus, Merismopedia, Microcystis,* and *Ankistrodesmus.* They found that *Microcystis* was preferred over other algae.

Experiments we have conducted clearly indicate saltwater species of algae like *Tetraselmis, Isochrysis*, and *Nannochloropsis* can be used with good results. Best results are obtained by allowing the algae to settle in the culture vessel, pour off the culture water and feed the algae concentrate. Commercial algae paste may also be used even though they are saltwater species. Hobbyists speak of "Delicious Daphnia Dreams" a home made food consisting of 8 ounces frozen peas, 3 ounces of carrots and/or beets, some paprika and 2 multivitamins. The ingredients are blended into a soup with a little water, placed into a fine net, squeezed, and the juice is retained and added to 1 gallon of water. This is then routinely added to the culture. Activated yeast is also widely used by hobbyist but report many failures due to over feeding.

Nutritional Value of Daphnia

Since a multiple of foods can be used to rear Daphnia the same question always remains, is the plankton animal nutritious and beneficial to the prey. The nutritional content of daphnia varies considerably depending on age and food type. Although variable, the protein content of daphnia usually averages 50% of the dry weight. Fat content (% dry weight) for adults and juveniles differ significantly, with a total fat content of 20-27% for adult females and 4-6% for juveniles (Ivleva, 1969). *Moina* grown in natural pond conditions have been reported to have 70% protein, 16.5% fat, 5% carbohydrates and 9.5% ash dry weight (Lee Heng, 1983). Whereas, *Moina* have also been reported to be 94.7% water, 3.9% protein, 0.54% fat, 0.67% carbohydrates, and 0.18% ash, with an energy content of 300 Kcal/Kg.

Fatty acid composition of food is important to survival and growth of intensively cultured fish fry. Omega-3 highly unsaturated fatty acids (ω3 HUFA) are essential for many species of fish. Fatty acid composition of *Moina* cultured on baker's yeast, freshwater *Chlorella*, and ω-yeast (yeast enriched with cuttlefish oil) has been compared (Oka, et al. 1982). *Moina* cultured with baker's yeast were high in monoenoic fatty acids such as 16:1 and 18:1. Those cultured with freshwater *Chlorella* were high in 18:2ω6 and 18:3ω3. *Moina* fed ω-yeast contained high concentrations of ω3 HUFA.

Experiments have also been conducted to improve the dietary value of *Moina* by direct enrichment with an emulsion of ω3 HUFA and fat-soluble vitamins (Watanabe, et al. 1983). *Moina* were found to take up lipids very easily from the emulsion and the concentration of ω3 HUFA in the *Moina* was proportional to the content of ω3 HUFA in the emulsion. The procedure for direct enrichment of *Moina* is identical to that described for rotifers (see Nutritional Value of Rotifers). Yet, a word of caution may be needed. Lee Heng (1983) stated that an increase of fatty acid levels in the diet of *Moina* reduces productivity. Therefore, fatty acid enhancement should be restricted to just prior to utilization.

Although little information is available on the nutritional value of hemoglobin levels we speculate that high levels may have positive effects in feeding larvae. Hemoglobin formation is dependent on the level of dissolved oxygen in the water. The production of hemoglobin may also be caused by high temperatures and high population densities (Rottman 1992).

Culture Methods

Culture vessels

Continuous daphnia cultures have been maintained in two liter cola bottles. However, these are usually too small to yield enough biomass to serve as fish food. We recommend at least a 38 liter (10 gal) aquarium. Shallow tanks with a large surface area are more suitable. Glass aquaria, tanks or vats (concrete, stainless steel, plastic, or fiberglass), and earthen ponds also can be used. Children's swimming pools, plastic sinks, bathtubs, and painted galvanized cattle watering troughs also work well. Do not use unpainted metal containers unless they are stainless steel.

Galvanized tanks can be painted with epoxy, but first they must be thoroughly washed with either acetic acid (vinegar) or copper sulfate solution to etch the galvanized surface so that paint will adhere. The tank should then be thoroughly washed with fresh water and completely air dried. The epoxy is applied in several thin coats rather than one thick one. A paint sprayer works better than a brush for this purpose. After the paint dries completely, the tank is again washed and can then be used to culture daphnia.

Location

Utilization of open pond cultures is discouraged since production can be highly variable. May, et al. (1984) suggest that if cultures are protected from rain, production is more consistent. In addition, light and temperature requirements are more easily controlled in a closed facility. Closed facilities also help reduce predation and contamination.

Water depth

Ivleva (1969) suggests that water depth should be less than 1 m (39"). Ventura and Enderez (1980) recommended a water depth of 0.4-0.5 m (16-20"). Shallow water depth with large surface area allows good light penetration for photosynthesis by microalgae and provides a large surface to volume ratio for oxygen diffusion. We have found that depths of 30-38 cm (12-15") are ideal, whereas deeper depths are less productive.

Aeration

Gentle aeration of daphnia pools increases algal production, the number of eggs per female, the proportion of egg-bearing females in a population, and population density (Heisig, 1977). Only one or two air lines are required in culture containers up to 1.5 m^3 (396 gal). Small bubbles should be avoided when culturing *Moina* since they can get trapped under their carapace, causing them to float to the surface and die.

Light and temperature

Ivleva (1969) recommend either diffuse light or shade over a third to half of the culture container surface. We found polyfilm covered greenhouses ideal, but heavy shade cloth (50 to 80%) should be located over central half or more of the house. Although greenhouses are ideal for lighting and heating, prolonged excessive heat during the summer months must be controlled. Automatic ventilation and/or removal of polyfilm along the lower quarters of the greenhouse helps control excess heat. *Moina* tollerate a wide range of temperature from 5-32°C (41-90°F) but the optimal temperature range is 24-31°C (75-88°F). Rottman (1992) states that they can easily tolerate a daily temperature variation from 41-88°F (5-31°C). Although *Moina* tolerate temperatures in excess of 32°C (90°F), reproduction and survival is quickly curtailed at sustained high temperatures. Diurnal fluctuations appear to be advantageous. Larger species, *Daphnia magna* and *pulex* have lower optimal temperature ranges from 18-22°C (64-77°F)

Water exchanges

In continuous cultures a small trickle of fresh water into the culture container may improve production of daphnia (Ivleva, 1969). A 25% water exchange has been suggested when continuous cultures show declining reproduction, but care should be taken not to greatly alter pH.

Culture Food Requirements

Microalgae nutrient sources

Whether using a continuous or batch culture method, utilization of microalgae produces the highest yields. Organic fertilizers are usually preferred to mineral fertilizers because they promote bacteria and fungal growth, detritus as well as nutrients for microalgae production. Normally these items would be considered contaminants, but in daphnia culture they constitute part of the "nutritional soup."

Fresh organic fertilizers are preferred to old because they are richer in organic matter and microbes. This is especially true of manure, which is usually dried before use. However, some farm animals are often fed a variety of prophylactic antibiotics and additives which may inhibit daphnia production. Drying or processing manures lessens the potency of these drugs. Commercially available manure such as dehydrated cow manure is suitable if fresh manure is unavailable. Dried, processed, sewage sludge (Milorganite) is an excellent and consistent nutrient source.

Mineral fertilizers may be used alone, however, they work better in earthen ponds rather than lined ponds tanks or vats. In ponds, the combined application of organic and mineral fertilizers creates favorable conditions for the growth of a variety of foods for daphnia. This variety of food items more completely meets the nutritional needs of the daphnia, resulting in maximum production. The combination of mineral fertilizer and yeast has been reported to improve production by a factor of 1.5-2.0 as compared to mineral fertilizer alone (Ivleva, 1969).

Utilization of manures and sludges

Organic nutrients (manures and sludge fertilizers) can be added to cultures in several ways.

1) Soaking the dry material for several hours and then distributing the moist material over the entire bottom, allowing it to slowly deteriorate.

2) Place the dry material in a mesh bag (tea bag approach) and suspend the bag inside the culture tank near an air supply for circulation and slow leaching.

3) Soaking the material for extended periods of time so that it decomposes into a concentrated "nutrient slurry" and add only the liquid to the cultures.

In each case the objective is to "digest" or breakdown particles into nutrient sources for enhancing bacteria and microalgae growth. Of the three methods, we prefer the last since it produces more consistent results at a higher range of temperatures, light, and water quality conditions. Nutrient, algae, and bacteria levels can be monitored to determine when new nutrients should be added.

Coarse organic materials such as manure, hay, bran, and oil seed meals are usually soaked in water and then poured into a bag that is suspended in the water column. Cheese cloth, burlap, muslin, nylon, or other relatively loose weave fabric may be used. Nylon stockings work well for this purpose. They are inexpensive, readily available, and nylon also holds up much better in water than cotton or burlap. This step is not absolutely necessary, however, it does prevent large particles from becoming a problem when the daphnia are harvested.

Although manure is widely used in commercial daphnia culture, for small-scale home aquarists, this nutrient source is objectionable especially in the home. Utilization of cultured microalgae (see Microalgae Section), yeast, or bran is more suitable.

Supplemental foods

Organic manures provide detrital food particles which are utilized directly as a food source, but manures mainly serve as a nutrient source for bacteria and microalgae. Activated or inactivated yeast is readily available and works well. Overfeeding is less of a problem with inactivated yeast. Inoculating cultures with microalgae improves water quality, reproduction, and provides essential nutrition to daphnia that is not found in yeast or yeast-based products. Micronized supplemental foods such as yeast (active or inactive), alfalfa, bran, wheat flour, dried blood, etc. can be provided. Overfeeding of these feeds can quickly cause fouling so care must be taken. Start with small amounts initially and slowly increase as demand increases. If fungal growths appear on the walls of the culture vessel, discard the culture. Bacteria promoted by deterioration of these foods and the organic nutrients are a major food item for daphnia, however, we have found that they are an inferior single food source compared to a combined diet of microalgae and inactivated yeast. For small cultures, Roti-Rich™ , a yeast based, fortified mixture, (Florida Aqua Farms) can be economically used in conjunction with microalgae. A good rule to remember is that yeast alone is not as nutritious and can easily be overfed leading to a crash of your cultures. Microalgae in conjunction with a yeast based food is considerably safer and will result in higher reproductive rates.

Contaminants

Contaminants are much more common in continuous and semi-continuous cultures than batch cultures. The longer a culture is maintained, the greater the chance of contamination. Rotifers, mosquitos, and desmids were the most common problems in continuous culture tanks. Hobbyists often report that Hydra are a common contaminant in sustained cultures. Whereas, in batch culture tanks rotifers were the most common problem. Elimination of rotifer resting eggs is difficult. If rotifer contamination becomes a serious problem, a good cleaning must be initiated which includes a good scrubbing, followed by heavy chlorination of the walls and bottom of the tank. If contamination persists, try cleaning with a 30% solution of muriatic acid.

Pond cultures have mosquitos, *Cyclops* copepods, diving beetles, dragonfly larvae and *Hydra* as common problems. However, filamentous algae can be especially troublesome in daphnia cultures. Webster and Peters (1978) reported that as the concentration of filamentous algae increases, the filtering rate of daphnia decreases, rejection rate increases, and brood size diminishes.

Chironomid larvae (blood worms) may occur in large numbers, especially in outdoor cultures, reducing the daphnia crop. These, however, can also be used as food for larger fish. Screening daphnia pools to prevent entry of flying insects helps control contaminating organisms.

Containers to be used for daphnia culture, whether aquaria, tanks, vats, or ponds, need not be particularly clean. However, for good steady production, consistent cleaning practices are required. Tanks can be disinfected by cleaning with a 30% solution of muriatic acid, followed by heavy chlorination, and drying in sunlight. Earthen ponds should be either sun-dried or treated with hydrated lime.

Daphnia inoculants

Adult females bearing eggs are preferred as inoculants since they are ready to spawn which reduces start-up time. This is particularly important when using batch cultures. Adults can be separated by using a larger mesh net (500 - 800 μm).

Avoid using animals from poor or declining cultures, those producing resting eggs, or cultures with contaminants such as rotifers. Best results are obtained using adults from fast growing cultures. Animals with high hemoglobin levels and good coloration are preferred. Inoculants should be washed to remove attached rotifers and other contaminants. Sometimes, a short 5 to 10 second dip in water high in ammonia, having high pH, or containing formalin followed by a good rinsing will help remove attached contaminants. Based on the fertilizer and food levels listed in Table 9.1, initial inoculant levels are recommended at 25/L (100/G). Inoculation densities should be approximately 3 *Daphnia* or 25 *Moina* per liter (10 *Daphnia* or 100 *Moina*). Inoculate cultures 24 hours or more after fertilization to allow for decomposition and development of bacteria and nutrients. Utilization of a predigested slurry as a nutrient source reduces inoculation time to a few hours. Inoculation levels are not rigid and can be altered depending on conditions. Ideally, an initial low mortality after stocking is normal, but if conditions are not suitable additional inoculations may be required. A greater inoculation density reduces the time to harvesting, but does not necessarily result in higher production. Concentrated semi-moist *Moina* are about 2,900 animals per 0.6 ml (1/8 teaspoon).

Cultures are usually inoculated 24 hours or more after fertilization. However, when yeast or cultured microalgae is used, daphnia can be added to the culture after only a few hours of aeration, assuming good water quality. This is because the yeast and microalgae cells are immediately available to daphnia as food.

Culture Methods

Continuous, Semi-continuous and Batch Cultures

A continuous culture means the same water is used for an extended time until conditions are no longer productive. In this case the fertilizer, microalgae, and micronized foods are added initially and replenished at 50 to 100% levels every 5 to 8 days. Duration of continuous cultures can vary due to environmental conditions, feed and feeding practices and contaminant levels. Usually they last for about 1 to 2 months before production is severely curtailed.

In semi-continuous cultures a portion of the water is removed at each harvest and new fertilized water and food are added. Using this process stabilizes the water quality and prolongs productivity. Duration of these cultures are usually 2 or more months.

Batch culture is short duration, where animals are either added to an existing microalgae culture and fed supplemental micronized foods daily, or they are added to plain culture water to which microalgae and micronized foods are added daily. In batch culture the same water is used for a specified time period (usually 5 to 10 days), everything is harvested, then new cultures are initiated. Batch cultures provide the highest yield and are better suited for projects that require a specific number of daphnia each day. In addition, these are purer cultures since there is less chance of contaminates becoming well established.

Food & Nutrient Source	At 375/L	At 100/L	Next Nutrient Dose
baker's yeast alone	8.5-14.2 g	0.3-0.5 oz	50-100% 5 days
ammonium nitrate baker's yeast	14.4 g 8.5-14 g daily	0.3-0.5 oz 0.5 oz	50-100% 5 days
alfalpha pellets or meal wheat or rice bran yeast	42.5 g 42.5 g 8.5 g daily	1.5 oz 1.5 oz 0.3 oz	50-100% 5 days
dried cow manure or sludge wheat or rice bran yeast	142 g 42.5 g 8.5 g daily	5.0 oz 1.5 oz 0.3 oz	50-100% 5 days
dried cow manure or sludge cotton seed meal yeast	142 g 42.5 g 8.5 g daily	5.0 oz 1.5 oz 0.3 oz	50-100% 5 days
dried cow + horse manure + sewage sludge	567 g	20 oz	50-100% 6-10 days
dried chicken or hog manure	170 g	6 oz	50-100% 5 days
dried sludge slurry blood meal	1/2 L 20 ml daily	16 fl oz 0.7 fl oz	50% 5 days

Table 9.1. Eight nutrient and food combinations used to raise continuous cultures of daphnia. Dosage is based on 375 L (100 G). Fifty to 100% of the initial dosage is usually added between days 5 and 8 to replenish the nutrient supply. Microalgae inoculants are added in each method. (From: Ivleva, 1969; Galstoff, et al 1937; Rottmann & Shireman, 1989; Hoff and Wilcox, 1986).

Day 1	Wash tank, spray with chlorine, allow to dry, rinse and fill to 12"
Day 2	Add 16 lbs dry sludge (Melorganite), 3 lbs cotton seed meal, 18 gallons of good microalage culture, aerate vigorously
Day 5-6	Allow nutrients to break down, bacteria and algae to grow
Day 7	Add 22 fl oz ammonium hydroxide to control contaminants
Day 8	Add 2 fl oz of washed adult Moina, reduce aeration
Day 8	Clean and start a second tank like the first
Day 9-12	Allow Moina to reproduce
Day 13	Begin harvesting 25% of the culture
Day 14-21	Harvest 25% daily
Day 15	Clean and start a third tank like the first
Day 22	Drain, clean and re-start first tank

Table 9.2. Continuous culture method used at VW Tropical Fish Farm (Hoff and Wilcox 1986)

Procedure for Continuous Culture

Daphnia can be continually cultured in combination with their food, or separate cultures can be maintained. Combined culture is the simplest, but production from separate cultures has been reported to be approximately 50% higher. Production from separate cultures has the disadvantage of requiring additional space for the cultivation of microalgae. Consequently, production per unit volume, especially in large scale cultures, is equivalent or even higher when the animals are grown together with their food (Ivleva, 1969). Nonetheless, there are distinct advantages of separate cultures of *Daphnia* and microalgae such as: 1) less chance of contamination, 2) greater degree of control, and 3) more consistent yield.

Table 9.1 lists eight combinations of nutrients and supplemental foods used to raise daphnia. Continuous daphnia culture using microalgae as food is similar to that described for rotifers (see Rotifer Culture). Microalgae is produced according to procedures presented in the Microalgae Chapter. In each case sufficient microalgae inoculants are added to produce a slightly green color, approximately 37L of inoculant per 375 L of culture (10 G/100 G) or less depending on algae density. In addition, an inoculant of adult daphnia are added to the culture water. The food and nutrient sources are not presented in any particular order, but are given as examples of materials that are easily available and work well. We suggest trying several of these to determine which works best in your situation.

Duration of a continuous or semi-continuous culture

Duration of continuous and semi-continuous cultures is directly proportional to the degree of maintenance, food and feeding, and water quality. Usually a culture will last 4 to 12 weeks before fouling or contaminants interfere with production. At some point cultures will fail to respond to additional fertilization. When it is evident that daphnia are not multiplying well, a new culture should be started. If the culture water is not loaded with contaminants, adding 5 to 10% of the old water to a new culture will help reestablish bacteria, fungi and yeast. Filtering this water through a 10μm screen will help reduce larger contaminants.

A shorter continuous method used in 1986 at VW Tropical Fish (Hoff and Wilcox 1986) is outlined in Table 9.2. In this procedure, three 3.7 x 2.4 x 0.5 M (12'x 8'x 21") deep tanks were used. Water depth was 30 cm (12") = 2,718 L (718 gal). This was a flexible program (± 4 days) based on the time of the year, temperatures, ambient light and *Moina* densities.

Once the culture was established, increasing amounts of inactivated yeast were added daily. To reduce sedimentation and anaerobic conditions, the tank bottoms were brushed daily. This practice suspends uneaten food particles and stimulates decomposition of organic fertilizer particles.

Harvesting

A partial harvest at regular daily intervals is necessary to maintain steady productivity of the culture. Initial harvests from continuous cultures should be not more than 1/5 to 1/4 of the population daily. Higher or lower harvesting may be required depending on the growth of the population. Daphnia can be harvested by simply dipping out the required number with a net or by siphoning into a net. A 150 μm to 200 μm net is suitable. When aeration is suspended, daphnia will concentrate on the surface and can be netted out. An alternative is to drain 1/5 to 1/4 of the water into a collector and replace the discarded water with new fertilized water (for semi-continuous cultures). For inoculants, a 500 to 800 μm net can be used to separate adults from young.

Procedure For Batch Culture

Batch culture of daphnia is nearly identical to the rotifer batch culture procedure (see Rotifer Chapter). Briefly, a microalgae culture is bloomed to a high density, pH is adjusted to between 7 and 8 with acetic acid, and then inoculated with adult daphnia. Supplemental feeding of Roti-Rich™ or a similar yeast-based product is provided in increasing levels as the cultured animals consume the microalgae. All algae is usually consumed within five days and the daphnia population is harvested. Batch culture is particularly useful when a specific quantity of daphnia is needed each day. Figure

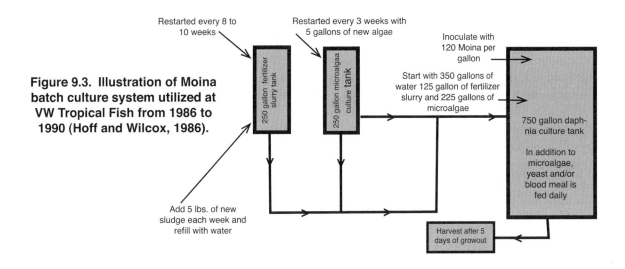

Figure 9.3. Illustration of Moina batch culture system utilized at VW Tropical Fish from 1986 to 1990 (Hoff and Wilcox, 1986).

Day 1	Add 125 gal of slurry to culture tank Add 220 gal of 5 day old microalgae culture Add 373 gal of water Add 120 Moina per gal	Refill slurry tank, add 5 lbs dry sludge Refill algae tank, add 5 gal slurry
Day 2	Add 38 cc yeast or 78 cc blood meal	Brush bottom of slurry tank Brush bottom of algae tank Brush bottom of Moina tank
Day 3	Add food as in day 2 or more if needed	Brush bottoms as in day 2
Day 4	Add food as in day 2 or more if needed Add ammonia or formalin to algae Add ammonia to Moina if needed	Brush bottoms as in day 2
Day 5	Add food as as in day 2 or more if needed	Brush bottoms as in day 2
Day 6	Harvest Moina Clean tank, wash with chlorine, dry Re-start tank as on day 1	Continue using algae tanks for 15-25 days, then re-start Continue using slurry tank for 8-10 weeks before re-starting

Table 9.3. A commercial batch culture method for Moina used over three years at VW Tropical Fish Farm (Hoff and Wilcox, 1986)

9.3 and Table 9.3 illustrate a *Moina* batch culture method used at VW Tropical Fish Farm from 1986 to 1990 (Hoff and Wilcox, 1986). The method was a three stage operation consisting of an organic slurry nutrient media tank, a microalgae growing tank, and the *Moina* culture tank.

Nutrient slurry tank

The organic fertilizer utilized was processed sludge (Milorganite). Initially about 7 Kg (15 lbs) of dried sludge was added to a standard 938 L (250 G) cement coffin vault, with moderate aeration and allowed to decompose into a nutrient slurry. The fertilizer tank was routinely re-stocked with 5 lbs of fertilizer in each growout cycle. Bottoms of the slurry tanks were brushed daily to suspend and aid digestion of fertilizer. After 8 to 10 weeks the slurry tank was drained, cleaned, and restarted.

Microalgae culture tank

Microalgae cultures were grown in a cement coffin vault under ambient lighting and strong aeration. Cultures were started using 19 L (5 G) sub-cultures that were maintained in another area. In addition to the microalgae, 19 L (5 G) of slurry and 832 L (220 G) of well water were added. After a 5 day grow out, all but 94 L (25 G) were drained into a 2714 L (718 G) daphnia culture tank. The microalgae tank was then re-fertilized with 19 L (5 G) of slurry and refilled with water. To control rotifers and other zooplankton during the algae culture period, 200 ml (7 fl oz) of ammonium hydroxide (29.4% active) was added. Formalin, which is more effective, was later used at 10 to 20 ppm. Care should be taken to add formalin a minimum of 24 hours prior to using the microalgae as feed. Microalgae tanks were re-bloomed 3 to 5 times before cleaning and starting a new culture.

Daphnia culture tank

Tanks were set up on a 5 day growout cycle. All but 25 gallons (11%) of an algae tank was transferred into a 2714 L (718 G) tank, 30 cm (12") deep. In addition, 473 L (125 G) of slurry and 1410 L (373 G) of water were added. Adult *Moina* were stocked at 120 animals per 3.78 L (1 G). During non-ideal conditions, higher inoculant levels were used to compensate for initial die-off. Yeast was fed as an additional food source at 38 cc (2.5 tbs) once or twice a day depending on age, microalgae density, and *Moina* density. Blood meal was later fed exclusively or in addition to yeast at a rate of 74 cc (5 tbs). In both cases, these foods were mixed with water prior to feeding. The bottom of the tanks were brushed once a day. Make sure a different brush is used for each culture tank and that it is clean each time. Aeration was moderate with 2 to 3 coarse bubble outlets per tank.

Yields

Using the batch method described above, the following averaged yields of *Moina* were obtained over a three year period (Table 9.4). This production data was from the harvest of 4 culture tanks each day with a total volume of 10.86 M^3 (2872 gal). Average production was 4.9 liters of condensed, drained *Moina* per day which is about 24 million individuals. Normal inoculant was around 32-50 adult *Moina* per liter, with average harvest of 2210 per liter in a five day growout .

Generally production was lower during the months of August, September, and October when higher temperatures occur in Florida. Cold weather effects were more evident in late December and January. Highest yields generally occurred during periods of medium light, warm balmy days, followed by cool nights more typical of spring (late February - March and early April) and fall (November and early December) in Central Florida.

Because of the great size variation in *Moina* populations (400-1000 μm), tests of acceptability should be made to compare newly hatched *Artemia* against *Moina*. We found that 30 ml (1 fl. oz.) of *Artemia* cysts (grade A) yielded about 47 ml (1.6 fl. oz.) of biomass. A one pound can of *Artemia* cysts has 828 ml (28 fl. oz.) of cysts and yields 1319 ml (45 fl. oz.) of nauplii biomass. To match the overall average daily production of *Moina* biomass (4.89 L) it would take approximately 3.7 one pound cans of *Artemia* cysts per day.

Month	1988	1989	1990	Average
January	120L 3.9 L/day	146 L 4.7 L/day	114 L 4.7 L/day	137 L 4.4 L/day
February	211 L 7.5 L/day	113 L 4.0 L/day	182 L 6.5 L/day	168 L L6.0 L/day
March	172 L 5.6 L/day	155 L 5.0 L/day	130 L 4.2 L/day	152 L 4.9 L/day
April	176 L 5.9 L/day	141 L 4.7 L/day	166 L 5.5 L/day	161 L 5.4 L/day
May	186 L 6.0 L/day	129 L 4.2 L/day	158 L 5.1 L/day	158 L 5.1 L/day
June	214 L 7.1 L/day	105 L 3.5 L/day	94 L 3.2 L/day	138 L 4.6 L/day
July	229 L 7.4 L/day	132 L 4.3 L/day	114 L 3.7 L/day	158 L 5.1 L/day
August	128 L 4.2 L/day	128 L 4.1 L/day	123 L 4.0 L/day	126 L 4.1 L/day
September	154 L 5.1 L/day	118 L 3.9 L/day	123 L 4.1 L/day	132 L 4.4 L/day
October	186 L 6.0 L/day	121 L 3.9 L/day	123 L 4.0 L/day	143 L 4.6 L/day
November	197 L 6.6 L/day	159 L 5.3 L/day	121 L 4.0 L/day	159 L 5.33 L/day
December	133 L 4.3 L/day	189 L 6.1 L/day	125 L 4.0 L/day	149 L 4.8 L/day

Table 9.4. Average monthly production of Moina from 10.9 M³ (2872 gal) of culture water using the batch method. Top number is the total liters of condensed Moina harvested and bottom number is daily harvest. Data from VW Tropical Fish (Hoff and Wilcox 1986).

Monitoring and Assessing Daphnia Populations

Cultures should be inspected daily to determine their health. The following observations should be made. A hand lens (8-10X) is useful for checking the density of daphnia, the age distribution of the daphnia cultures (rapidly reproducing cultures have a large percentage of young), and to determine if rotifers or *Cyclops* have invaded the culture.

1) The population density of daphnia is determined by first stirring the culture and removing 10-100 ml, killing the daphnia with a 5% formalin solution and counting in a petri dish with a 8-10X hand lens or dissecting microscope. With experience, population density can be estimated visually without the need for daily counts.

2) When examined in a clear beaker, culture water should appear slightly cloudy, tea-colored, or green. Feed or fertilize whenever the transparency is greater than 40 cm (16"). This can be determined with a Density Measuring Stick™ (see Microalgae Chapter - Estimating Algae Cell Densities).

3) Green or brown-red daphnia with full intestinal tracts and active movement indicate satisfactory conditions. Daphnia clouds on the surface, pale colored daphnia, or daphnia producing ephippia may indicate sub-optimal conditions.

4) Cultures that do not produce maximum densities of 2,000 *Moina* per liter (8,000/gal) in 8-10 days or 25 *Daphnia* per liter (100/gal) in 16-26 days, should be completely harvested and re-started.

5) If rotifers or *Cyclops* are observed in high densities and the contaminants cannot be controlled by ammonia or formalin treatment, harvest the culture, clean the tank thoroughly, and re-start.

Cost effectiveness

Since *Moina* are considered a substitute for *Artemia* in the freshwater tropical fish industry, a cost analysis to raise *Moina* verses *Artemia* is provided based on work done at VW Tropical Fish Farm. Providing ponds or tanks are available, the cost to rear *Moina* is less expensive per unit volume of biomass (Table 9.4). Although there are more animals per unit volume in *Artemia,* it takes more nauplii to match the volume of one *Moina.* Labor is based on $8.00 /hr and more current prices for *Artemia* cysts.

Additional Points

Daphnia cultures tend to have more stable population densities than rotifer cultures. However, as with rotifers, it may not always be possible to match daphnia production to feeding demands of fish fry when using a continuous culture Harvested daphnia can be kept alive for several days in clean water in a refrigerator. They will resume normal activity upon warming. The nutritional quality of refrigerated

Artemia			
Labor	1,5 h/day	$8.00/hr	$12.00/day
Artemia cysts	3.7 cans/day	$45.00/can	$166.50/day
Miscellaneous	electric, salt	$6.00	$6.00/day
Cost per liter	based on 4.9 L/day	-----	$37.65/L
Moina			
Labor	4 hr/day	$8.00/hr	$32.00/day
Moina	$0.00	$0.00	$0.00
Miscellaneous	electric fertilizer & food	$10.00	$10.00
Cost per liter	based on 4.9 L/day	------	$8.57/L

Table 9.5. Comparison of costs between culturing the same daily volume of Moina compared to the same volume of newly hatched Artemia. Cost analysis is based on three years of data supplied in Table 9.4. Prices were updated on 4/23/01

daphnia will not be optimal because of starvation during storage, so daphnia should be enriched with algae, yeast, or a fatty acid emulsion before feeding to fish. Daphnia can be stored for long periods by freezing in low salinity water (7 ppt, 1.0046 density) or by freeze-drying. Both methods kill the daphnia, so adequate circulation is required to keep them in suspension after thawing. As with rotifers, frozen and freeze-dried daphnia are not as nutritious as live animals and they are not as readily accepted by fry. Although freezing or freeze-drying does not significantly alter the nutritional content of daphnia, nutrients leach out rapidly into the water. Nearly all of the enzyme activity is lost within ten minutes after introduction into freshwater. After one hour, all of the free amino acids and many of the bound amino acids are lost (Grabner, *et al*. 1981).

The surface area of a culture tank appears to have a positive effect on the production of daphnia. A four-fold increase of surface area, in the form of suspended plastic sheets, has been shown to result in a four-fold increase in population density and daphnia biomass harvested. It is unknown whether this is the result of improved water quality due to nitrifying bacteria on the substrata, a change in the spatial distribution of the daphnia, or improved nutrition (Langis, *et al*. 1988).

Daphnia are found throughout the world and as a result, differences in size, brood production, and optimal environmental conditions exist between different species and strains. Adjustments will need to be made in culture technique depending on the particular daphnia you wish to produce.

Trouble Shooting

Daphnia cultures occasionally fail or do not develop as expected. Some possible reasons for failure are listed below:

1) Toxic materials in the water supply - *Daphnia* are extremely sensitive to pesticides, metals, detergents and bleaches. A ion imbalance or over-fertilization with mineral fertilizers can also be toxic to *Daphnia*. Try an alternate water source and insure that toxicants are not inadvertently introduced into the cultures.

2) Temperature outside the optimum range - *Daphnia* species are especially sensitive to temperatures above 22°C (72°F). If your culture conditions exceed this value, use *Moina* which thrive at temperatures up to 32°C (90°F). Low temperatures also reduce *Daphnia* production.

3) Insufficient dissolved oxygen - Both *Daphnia* and *Moina* can survive with little dissolved oxygen. However, dense cultures should be aerated to provide sufficient oxygen. Culture tanks with high surface areas in relation to shallow depth are preferred.

4) Heavy aeration or small bubbles can strip the daphnia from the culture.

5) Too much food or fertilizer - overfeeding can quickly cause problems with water quality. Always start with small quantities and increase amounts as you gain experience. Excessive fungus growth in large quantities in the culture is an indication of overfeeding. If this occurs, the culture should be discarded.

6) High pH due to a dense algae bloom may result in an increase of un-ionized ammonia which may inhibit daphnia production.

7) Insufficient food or fertilizer - Complete clearing of the culture water is a warning that the daphnia will soon starve. Pale daphnia with empty digestive tracts or daphnia producing ephippia are also indications of insufficient food.

8) Insufficient inoculation density - Although a culture can theoretically be started with a single female, always use the recommended inoculation density to produce a harvestable population quickly.

9) Competition from other organisms - If rotifers, ciliates, or copepods are found in large quantities, harvest the culture and clean the tank to prevent contamination of other cultures.

10) Hydra are often a common contaminant in cultures obtained from wild sources and can completely distroy a culture. Use of natural pond water is not advised unless sterilized in a microwave since many contaminents have resting stage cysts which are hard to remove.

11) Rotifers are generally more sensitive to ammonia than daphnia. Consequently, rotifers can be suppressed in daphnia cultures by adding enough ammonia to raise unionized ammonia (NH_3) levels to about 20 ppm (Lincoln, et al. 1983).

12) When setting up another culture, better results are obtained when a 10% to 20% portion of the old water is transferred into the new culture vessel.

Chapter 10 - CLAM & OYSTER VELIGERS

Introduction

Substantial interest has developed using newly hatched, naked (without shell), motile, trochophores from clams and oysters as food for marine finfish larvae. Currently, several companies are shipping cyrogenically preserved oyster trochophores that begin swimming upon defrosting and are ready to feed to larval predators. Timing is of the essence when using these live food organisms especially at the elevated temperatures normally associated with tropical fish rearing. At 30°C (86°F) it may be only 8 hours until they metamorphosis into shelled veligers. Therefore, they must be kept cool to retard metamorphosis and fed to the larvae intermittently. If properly maintained, they metamorphose into several motile veliger stages before settling out of the water column. Normal life span of trochophores is around 24-36 hours. Even though veligers have a thin, transparent, shell, use of these motile stages should only be utilized for rearing fish, crabs, shrimp, and starfish equipped with crushing plates or the ability to consume a whole veliger.

The reason that veligers can be used as larval feeds is due to the prolific reproductive capacity of marine bivalves and urchins. For example, the reproductive capacity of the common Florida oyster *Crassostrea virginica* is 25 to 100 million eggs per spawn. In Florida, oysters spawn naturally several times per year. Fecundity and timing of spawning is directly related to the fat reserves which can be controlled or enhanced under culture. Natural spawning occurs in early spring, midsummer (June and July), and fall (September and October). In captivity oysters and clams can be induced to spawn by a rapid change in temperature. Males and females are usually placed in shallow troughs and forced to spawn by physiological and environmental manipulation.

Table 10.1 outlines the nutritional values for three adult common shellfish. However, little is known about the relative nutritional values for veligers and the range of fish or invertebrates that that consume them. Consequently, use of veligers as a live larval feed should be regarded at this point as basically experimental. Our discussion will concentrate on controlled spawning and early larval rearing of hard shell, quahog marine clams.

	Clams	Oysters	Scallops
Calories	50	80	100
Fat Calories	10	20	10
Total Fat	1 g	2 g	1 g
Saturated Fat	0 g	0.5 g	0 g
Cholesterol	45 mg	55 mg	40 mg
Sodium	65 mg	190 mg	185 mg
Total Carbohydrates	3 g	4 g	3 g
Proteins	10 g	9 g	18 g
Vitamin A	0 %DV	0 %DV	0 %DV
Vitamin C	0 %DV	0 %DV	0 %DV
Calcium	8 %DV	10 %DV	2 %DV
Iron	20 %DV	45 %DV	2 %DV
Omega 3 Fatty Acid	0.15 g	0.61 g	0.20 g

Table 10.1 - Shellfish nutritional values for 4 ounces (114 grams) of raw, edible portions

A considerable portion of these instructions come from an excellent clam culture manual by Hadley, et al. (1997). Our interest in this information is only in spawning and rearing of clams while in the motile stages for use as a larval feed. For those interested in rearing clams to market size, we strongly suggest obtaining a copy for reference.

Life Cycle of Clams

Clams are benthic and are termed infauna that are normally buried with their siphons extending to the surface of the substrate, yet their larvae are planktonic. Clams are mostly marine and usually found in water of 20 ppt (1.0145 density) or higher, except for some coastal river types which tolerate lower ranges. Generally they are dioecious (male and female), however, hermaphroditism (both sexes) and protandry (sex reversal) does occur (Huner and Brown, 1985). This chapter will refer to the marine quahog clam, the inshore variety *Mercenaria mercenaria* which is protandric, maturing as males at an early age and changing sex in subsequent years to spawn as females (Hadley, et al. 1997). In nature, the life cycle consists of an external release of sperm and ova (eggs), collectively called gametes, directly into the water column. Synchronous timing of gamete release is essential as both males and females are triggered by environmental cues. As discussed in the microalgae section, the spring bloom of microalgae coupled with seasonal water temperature rise or fall constitute the sexual cues that induce spawning. Beside temperature cues, release of gametes into the water column is detected by others within the population, thus triggering a mass spawning.

The marine quahog clam *Mercenaria mercenaria* can reach sexual maturity in less than a year and measure only 20 mm (0.8 in) in anterior-posterior length and have a fecundity of around 100,000 eggs (Eversole, et al. 1980). However, most successfully induced clams are about 2 years old and 50 mm in length and have a fecundity of 1 to 2 million eggs (Menzel, 1971). Large quahog clams, 4-5", spawn 3 to 5 million eggs/spawn. Warm water quahog clams can live over 25 years, whereas the cold water species *Arctica* can live over 100 years (Thompson, et al. 1980). On the east coast of Florida spawning generally extends from early March through May followed by another period from October through November. Sexual maturity cannot be determined externally so a fairly large broodstock (10+ individuals) may have to be conditioned with the hope that males and females are present in the population. However, due to the protandrous nature (males later become females) of *M. mercenaria*, larger individuals are probably females and smaller ones are males. If stock is being taken directly from the wild, then historical spawning information for that particular area and species may be helpful to determine the timing of spawning of that population. Stocks should be obtained a minimum of 2 to 8 weeks prior to needing a spawn.

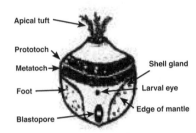

Figure 10.1 - Typical trochophore larvae

Shellfish in general produce a prodigious quantity of larvae per individual. A single medium size (2") hard clam or quahog (*Mercenaria mercenaria*) can conservatively spawn well over a million eggs. At fertilization zygotes (fertilized eggs) develop over a 24 hour period (at 28°C) to the initial free swimming trochophore (Figure 10.1) followed by the veliger stages (Figure10.3 & 10.4). Developmental stages include the early and late umbo stage, followed by the pediveliger stage, then into a post-set stage where the clam seeks a soft mud bottom. In this intermediate pediveliger stage sticky byssal threads are developed which help then to cling to surfaces. This short attachment stage is quickly followed by complete metamorphosis as they lose all swimming capability and "set."

Early, planktonic, motile stages (veligers) last from 8 to 21 days depending on environmental conditions. At 28°C they settle to the bottom in 8-10 days. At lower temperatures (20°C / 68°F) and with limited algal rations metamorphosis can be extended to 21 days. Veligers can be used as a small (60 μm), motile, live food. At the setting stage they are about 200 to 240 μm in size. Figure 10.2 below shows the basic anatomy of clams

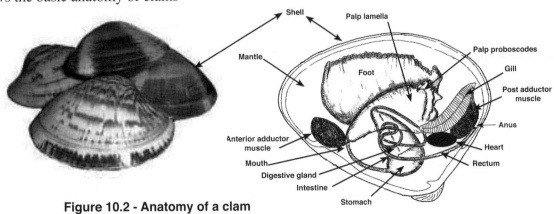

**Figure 10.2 - Anatomy of a clam
(modified after Barnes and Yonge)**

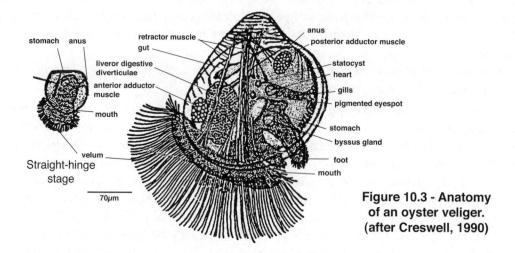

Figure 10.3 - Anatomy of an oyster veliger. (after Creswell, 1990)

Straight-hinge stage

stomach anus
retractor muscle
gut
liveror digestive diverticulae
anterior adductor muscle
mouth
velum
70µm

anus
posterior adductor muscle
statocyst
heart
gills
pigmented eyespot
stomach
byssus gland
foot
mouth

Life Cycle of Oysters

Compared to clams, oysters are extremely tough mollusks. Oysters are more euryhaline (capable of tolerating a wide range of salinities) than clams, tolerating salinities from 5 to 35 ppt (1.0030 to 1.0260 density). However, they prefer and proliferate in intertidal and subtidal beds in low salinity estuaries. Oysters are also eurythermal (capable of tolerating a wide range of temperatures) including direct exposure to hot summer and freezing winter temperatures. Intertidal oysters are exposed to desiccation for several hours without apparent physiological stress (Crestwell, et al. 1990).

The American oyster, *Crassostrea virginica*, is a protandrous hermaphrodite, first maturing as a sperm producing male, and later as an egg (ova) bearing female. As opposed to clams, oysters have a higher fecundity ranging from 25 million for a 3.5-5" female to 100 million for larger females. Spawning commences whenever water temperatures are above 25°C in Florida and will continue as long as the temperature remains above that minimum (Bardach, et al. 1972). In temperate waters, they spawn 1 or 2 times a year, while in subtropical waters, they are in a spawning condition 3+ times over the year. In Florida, peaks are in early spring (February through March), midsummer (June and July) and fall (September and October).

Spawning is synchronous and external, and like clams is triggered by warming waters in the spring and early summer or cooling waters in the fall. Like clams, the presence of sperm and eggs in the water column stimulates others to spawn as they siphon in the gametes during normal feeding processes. Oyster eggs are smaller than clam eggs, ranging from 35 to 45 μm, whereas clam eggs are 60 to 70 μm diameter.

Within minutes, fertilized eggs begin cell division. Over a period of several hours they will undergo repeated cell divisions and develop into an intermediary, free-swimming, stage called a trochophore (Figures 10.3 and 10.4). This ciliated stage is most often recognized by its swirling, swimming activity when examined under a microscope. Within several hours the trochophore transforms into the straight hinge or "D-staged" or "straight hinged" larvae. Larval development continues for about 16-20 days depending on temperature, food, and water

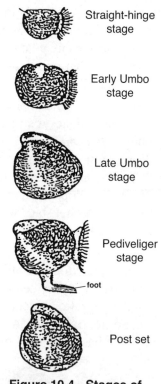

Straight-hinge stage

Early Umbo stage

Late Umbo stage

Pediveliger stage
foot

Post set

Figure 10.4 - Stages of larval development. (after Vaughan, 1987)

quality. At pre-metamorphose the oyster larvae is about 300 μm long and extends its newly developed foot in search of a suitable substrate for attachment. At this stage both the muscular foot, cement gland, and the photosensitive pigmented organ called an "eye" are fully developed. The pigmented "eye" helps the metamorphosing larvae to attach and glue itself into position. After finding a suitable permanent hard surface it will deposit a drop of "cement" and attach by its left valve. During this metamorphosis the oyster loses its swimming apparatus (velum) and begins to develop gills and sensing appendages called palps (Crestwell, et al. 1990).

Conditioning For Spawning

Conditioning for clams and oysters is basically the same. Two sets of tanks are normally used for broodstock holding/conditioning and spawning. Conditioning tanks are usually shallow troughs less than 20" deep. For small scale, tanks with high surface area and insulated side and bottom will suffice. Spawning tanks are usually shallow (less than 6" deep), narrow, and long, providing high surface area. They are also insulated and painted black for contrast of sperm and eggs. Stocking density for large scale operations is normally 4 liters (1 gallon) of water per clam. Ambient room temperature should be 24-28°C (75-82°F). Most hatcheries maintain broodstock at a cool 22°C (72°F). A recirculating or drop-in chiller will be necessary to maintain temperatures. Normally a 1/2hp chiller is sufficient to cool 1200 liters (300 gallons) of water to 20°C from an ambient of 28°C. Conditioning takes a few to several weeks if the broodstock has just spawned. If conditioned properly, broodstock can remain in pre-spawning state for up to six months.

Conditioning Temperature

Two types of conditioning regimes are commonly used. Long term consists of maintaining broodstock in cool 18-20°C water, feeding significant amounts of microalgae. Short term or "rapid conditioning" consists of maintaining water temperature at 22-23°C (72-73°F). Maintaining constant water temperature is critical with no more than a 1°C change. Higher temperatures are sometimes employed as a "priming" stage for a few days immediately prior to spawning. However, temperature control becomes more critical as you approach the spawning temperatures of 30°C (Hadley, et al. 1997).

Food and Feeding

Many algal species can be fed to broodstock. A varied diet is preferred since any single species may not be nutritionally complete (Hadley, et al. 1997). Most hatcheries utilize the naked flagellate, *Isochrysis galbana* (T-Iso) and the diatom, *Chaetoceros spp.* Feed requirements can be significant depending on physiological and environmental conditions. Clams typically consume $1\text{-}3 \times 10^9$ (1-3 billion) algal cells/clam/day. However, an algae density of 750,000 cells/ml should not be exceeded since it would actually inhibit feeding. At high cell densities clams produce copious, undigested, pseudofeces of undigested algae (Hadley, et al. 1997). Feeding dosage in batches or by constant drip or minute flow must be employed. The initial feeding should be about 100,000 cells/ml followed by lower densities of algae slowly added over a 24 hour period. Initially algal counts maybe required using a hemacytometer (see Algae ID Section), but over time an experienced culturist can determine cell densities by transparency. A fluorometer can also be used which provides a quick means of checking algal densities. Broodstock clams should be checked regularly, making sure siphons are extended (indicating feeding) and fecal material is present.

Although not currently mentioned by contemporary aquaculturists, considerable work was done by Ingle, et al. (1969) utilizing micronized cornmeal and other grains to fatten oysters. This work was conducted to enhance the taste and plumpness of oysters for human consumption by increasing the glycogen content. However, it may have application in a simple inexpensive way to condition oysters and clams.

Maintenance

If broodstock were obtained from natural conditions they should be thoroughly cleaned of all epiphytes and fouling organisms prior to introduction into the conditioning tanks. Two conditioning tanks are preferred so that clams can be quickly transferred (within 15 to 30 minutes) from one to the other during cleaning. Clams or oysters can be placed on plastic trays to expedite removal and cleaning. Conditioning tanks should be routinely serviced and cleaned by spot siphoning of fecal deposits. Depending on water quality and detritus conditions, once every one to two weeks clams should be removed from the tank and transferred into the reserve tank. Care should be taken to handle clams gently during the process since shock can trigger spawning in ripe individuals. Prior to placing them into the reserve tank, lightly wash tray and clams with cool tap water to remove excess silt, feces, and pseudofeces. Scrub and wash the conditioning tank with dilute bleach (10 ml household bleach per 4 liters). After bleaching, rinse thoroughly, refill with pre-chilled water and treat with sodium thiosulfate to remove any residual chlorine. If the tank is not being used immediately cover the top with black plastic to reduce fouling and keep the flow going. It also is suggested that the chiller and plumbing be flushed with chlorine. This can be accomplished by filling one of the empty tanks with chlorinated freshwater, turning off tanks containing clams, then recirculating the water through the chiller for 30 minutes. Add sodium thiosulfate to neutralize the chlorine and continue circulation for about 5 minutes, check for residual chlorine with a test strip and then flush out treated water. Open valves to tanks containing clams and resume circulation. Note that some maintenance can be reduced by using porous broodstock holding racks and sloped or "V" shaped bottom tanks so that fecal material can be simply flushed off the mollusks using a hose attached to the recirculating system. Turn off circulation to allow fecal material to settle, then siphon or draw off accumulated detritus.

Determining Sexual Maturity

Depending on the time of the year and initial state of broodstock, conditioning should be sufficient after 2 to 8 weeks. However, sacrifice of selected clams may be necessary to determine ripeness in order evaluate the current conditioning regime or determination of the condition of wild stock during normal spawning seasons. Care should be taken not to damage the body. A ripe male clam will have a creamy white gonadal mass covering the gut area whereas a ripening female will have a network of white tissue in this area, looking like white veins (Hadley, et al 1997). Puncturing or slicing the ripe tissue will cause the sperm to free flow. Males ripen sooner and it is often desirable to examine at least one female. Macroscopic examination of female gonadal tissue should reveal the tubules of the female's ovaries. However, microscopic examination may be required. A smear of gonadal tissue is placed on a slide with some warm seawater. Mature ova will be well rounded and uniformly opaque. Less mature ova may be teardrop shaped, stalked, and/or irregularly shaped with a transparent area. The larger the transparent area the less mature the egg. Mature sperm often need to be diluted with warm seawater to induce hydration and check mobility. For small scale aquaculture, sacrifice is probably not necessary, provided the broodstock has been fed and cared for under a strict temperature regime.

Preparation for Spawning

In preparation for spawning, remove conditioned broodstock, brush clean the shell, and submerge in a dilute chlorine bath (1 ml/liter) made with freshwater, not saltwater, for 5 to 10 minutes, then rinse well. Place cleaned clams in a refrigerator or insulated cooler with gel-packs. Damp towels, or newspapers should be placed around the clams to prevent dehydration.

Spawning Procedure

Cleaned mollusks are induced to spawn by using one or more thermal shock cycles consisting of raising water temperatures from 20°C (68°F) to about 30°C (86°F) in 20 minutes. If spawning does not begin within 30 minutes, the temperature is rapidly dropped to 20°C and the cycle is repeated. Preconditioned clams are placed shallow troughs or glass baking dishes which are slowly flooded with warm seawater followed by cool seawater. For small, more controlled conditions a water bath is preferred. One or two clams are placed in a small volume (±1 liter) glass or plastic container in cool saltwater (20°C/68°F) submersed in a water bath. Allow clams to acclimate and begin siphoning before raising the temperature. Flood water bath with hot tap water and recirculate with a small submersible pump to provide even warming. The objective is to raise the temperature to 28-30°C (83°-86°F) with a maximum of 32°C (90°F) over a 20-30 minute period. Once the desired temperature has been reached, maintain for a maximum of 30 minutes unless the clams appear stressed (noted by cessation of siphoning and closure of the shell valves). After 30 minutes rapidly cool the clams to 20-22°C (68°-72°F) hold for 20-30 minutes and repeat the process until spawning commences. In either spawning regime it is imperative that the water in which the clams are retained is clean and prefiltered through a 1 μm to 5 μm particle filter.

If spawning does not commence, further stimulation can be achieved by sacrificing a conditioned clam and using its gametes to trigger spawning. Make several slices through the gonad, then wash sperm or eggs into a beaker of seawater. Sperm can be used immediately or preserved for future applications. If preserved, heat the solution to 60°C for 10 minutes, cool, check under a scope for activity, and store. Pasteurized clam sperm can be retained in the refrigerator for several days prior to use. If the sacrificed clam is a female, wash eggs into a beaker then strain from the solution using a 20-35 μm screen into a second beaker. Use the sperm or egg wash (without the eggs) as the stimulus. Using a small pipette or eye dropper slowly introduce a small stream of the solution near the incurrent siphon of each clam. Sperm or egg extract is ingested through the incurrent siphon providing a strong stimulus for spawning. Rarely does this method not elicit spawning (A.C.T.E.D., 1997) in properly conditioned clams. Note that this stimulus induces clams or oysters to spawn in the warm stage not the cool stages of the temperature cycles.

Detection of spawning is done visually. Sperm mixes quickly with water and appears as a whitish or "smoky" color cloud. Eggs do not disperse quickly and impart a granular or small particle appearance in the water. Often they settle to the bottom in mounds.

Ideally fertilization should be accomplished within 30 minutes of spawning. However, one of the problems in forced spawning is that it is not always synchronized. Males might spawn quickly, but females may take more time. Unfertilized gametes deteriorate rapidly and measures should be taken to keep them in a viable condition until fertilization can be completed. Eggs are best held at room temperature, while sperm should be chilled. However, the longer the gametes are held, the lower the fertilization rate. After 2 hours the viability of eggs is 50% at 22°C and 30% at 4°C while the viability of sperm is 55% at 22°C and 40% at 4°C. However, the viability of eggs at 5 hours is 35% at 22°C and 0% at 4°C, while sperm is 0% at 22°C and 35% at 4°C (Hadley, et al. 1997). If retained longer than an hour prior to fertilization, eggs should be screened onto a 20-35 μm screen hourly and placed in new filtered, temperature adjusted water.

If using a mass spawning trough where the clams or oysters are all retained in the same water it will be necessary to remove spawning males quickly and put them in a separate common aquarium or bucket with slow aeration. The purpose of this is to reduce the amount of sperm in contact with the eggs. Excessive sperm results in multiple penetrations of the eggs causing "polyspermy", a condition that leads to abnormal and aborted development of eggs (A.C.T.E.D., 1997). Just 1 to 2 cc of moderately concentrated, viable sperm solution will sufficiently fertilize several million eggs.

Spawning females should also be pooled into an aquarium to concentrate eggs. When spawning commences those individuals should be removed from the spawning table. If using one clam/container in a water bath, set it aside or pool into a larger container. Spawning may continue for up to an hour. If retaining broodstock for future spawns, remove from the spawning container and mark as to sex and date using a waterproof marking pen. Place post-spawning broodstock back into conditioning tanks.

If spawning is conducted in a common container, a percentage of the eggs are probably already fertilized regardless of your attempt to quickly remove the males. However, if the spawners have been isolated in individual containers, it will be necessary to fertilize eggs by adding sperm. Pooled eggs should be diluted to about 200/ml if retained over 30 minutes prior to fertilization. High density fertilization of 2000 eggs/ml is also used, but eggs can only be retained for about 30 minutes. Add sperm to the pooled eggs at a rate of 1000-2000 sperm for each egg. Determination of sperm concentration is difficult due to their high concentration and small microscopic size. If sperm concentration is unknown, add in small units and microscopically examine eggs for fertilization. After 5-10 minutes check a 1 ml sample at 100X. If fertilized, one or more sperm may be observed clinging to the gelatinous sheath surrounding the egg. If fertilization occurred more than 15 minutes prior to examination, the first polar body may be apparent, appearing as a small bump protruding from the side of the egg. If fertilization is not apparent within 15-20 minutes, add another aliquot of sperm and observe in 5-10 minutes (Hadley, et al. 1997).

Within an hour or less, eggs should be sieved onto a 20-25μm screen and rinsed gently with clean seawater to remove excess sperm. Then transfer into a new clean container with clean filtered seawater. Unfertilized eggs will quickly deteriorate and several washings may be necessary over the next few hours to fully clean the fertilized eggs (zygotes).

Larval Rearing

If the decision is to take the non-feeding trochophores into the veliger stages, larval rearing equipment will have to be in place.

Tanks and stocking densities

Larval rearing tanks can range from 1 liter beakers to 12,000 liter tanks, depending on what level you desire. The main consideration is good water quality, and elimination of competitors and contaminants. Static, conical bottom or slope bottom containers with central or end drains are the easiest to drain and clean. Central recirculating systems cannot be used because the primary food is microalgae and it would be removed using normal filtration systems. Static tanks equipped with moderate aeration are used with periodic water exchanges and algae additions. Normal stocking density of fertilized eggs (zygotes) is 20-30/ml or 2 to 3 million per 100 liters. Often downwellers are used, which are tanks with screen bottoms that float or are positioned along the sides of a large circular or rectangular tank. These are equipped with an outside airlift pump that removes water from the larger tank and deposits it in the larval rearing tank. This water then flows back out of the bottom of the larval tank. Algal concentrations are maintained in the larger tank.

Larval rearing

After several hours the fertilized zygotes (eggs) enter into the naked, motile trochophore stage which lasts for about 24 hours then they enter into a shelled veliger stage which lasts from 8 to 21 days, but 10-12 days is average. Survival rates to this stage can vary from 0-100% with 25-90%

Stage	Age	Consumption of Algae Cells /Clam/Day	Food Concentration Minimum	Algae Cells/ml Maximum
Early veliger	1-2 days	1,000 - 5,000	10,000 - 25,000	50,000 - 75,000
Early veliger	3-5 days	5,000 - 10,000	same	same
Mid veliger	5-8 days	10,000 - 15,000	same	same
Late veliger	8-14 days	15,000 - 30,000	same	same
Pedi veliger	14-21 days	30,000 - 50,000	same	same

Table 10.2 - Recommended feeding concentrations of *Isochrysis galbana* to clam veligers
(after Hadley, et al. 1997)

Days at 28°C	Size Standard Length	Bottom Screen Mesh	Larval Density/ml	Algal Density Cells/ml
2	70-80 μm	35 μm	5/ml	20,000 - 25,000
4	120 μm	53 μm	5/m/	20,000 - 25,000
6	180 μm	53 μm	4-5/ml	30,000 - 40,000
8	250 μm	75 μm	4/ml	50,000
10	280 μm	75 μm	4/ml	50,000
12	300 μm	100 μm	3/ml	70,000 - 80,000
14	320 μm	125 μm	2-4/ml	100,000 - 150,000

Table 10.3 - Recommended feeding concentrations of algae per ml for oyster veligers.
Screen indicated is "plankton cloth" that is glued on the bottom of downwellers.
(after Creswell, et al. 1990)

being typical. During the trochophore stage the tank is filled with warm (24-28°C/75-83°F), filtered water, no microalgae is added and little to no aeration is provided.

Healthy veligers will actively swim and spend a large proportion of their time near the surface. Allow 6 hours before handling these early stage veligers to allow the shell to harden sufficiently. Siphon off the healthy veligers onto a 45 to 55μm screen. Note, less healthy and dying individuals will be in lower 1-3" bottom region of the water column and are not collected. Gently rinse the collected veligers with saltwater and stock into a clean tank at 10-20/L. Feeding must commence once the distinctive, straight hinged, "D" shaped veliger is present. Enough algae must be added as shown in Table 10.2. This table clearly outlines the stocking densities and algae densities needed (Hadley, et al. 1997). Exchanges using 20-25ppt salinity, temperature adjusted and filtered to 1 to 5μm water should be made daily to three times a week. Table 10.3 by Creswell, et al. (1990) was developed for feeding oyster veligers at 28°C (79°F).

Feeding assessments can be made by observing how quickly algae is removed from the water. A reduction in clearing time may indicate poor health or mortality. Pink areas on the walls or bottom of the tanks indicate high levels of sulfur bacteria which suggest poor feeding or dirty tanks. Siphoning and water exchanges may have to be increased. Affected containers and delivery lines should be cleaned and chlorinated. A slimy film on the walls of the tanks often indicates overfeeding.

At 180-200μm the late umbo stage larvae begin metamorphosis into the pediveliger stage. This is the last stage before the mollusk becomes benthic for the rest of its life. Like many aquatic animals, mortalities are often high at metamorphosis and good water quality is critical. Open flow downwellers or upwellers are often used during this last stage.

Feeding

High algae concentrations are counter productive. Algae densities should be 10,000 - 25,000 cells/ml. In addition, the amount of algae allocated for each larvae per day must be added in increasing densities depending on veliger age and density. Algae is added in either batches or continuously over a 24 hour period. Each batch addition should result in a density of 25 to 50 thousand cells/ml. Two or three batches will probably be needed daily. Stock algae cultures of 5 million cells/ml or higher are preferred to reduce bacterial contamination via the algal culture. Verification of algae density is made with a hemacytometer, fluorometer or "density stick" (available from Florida Aqua Farms).

Initially for a period of 7 to 10 days depending on growth stages, feed small flagellated algae like *Isochrysis*. Towards the end of the larval period the diatom *Chaetoceros* is added in addition to *Isochrysis*. Initial mixture is 25% of the daily ration gradually increasing to 50% as they reach metamorphosis. Since *Chaetoceros* is approximately 1.5 times larger than *Isochrysis,* the total cell density can be decreased proportionally. For example, if the *Isochrysis* density needed was 60,000 cells/ml, a 50:50 ratio would be 30,000 cell/ml T-Iso. and 20,000 cell/ml of *Chaetoceros* (30,000 divided by 1.5) (Hadley, et al. 1997).

Trouble Shooting

Heavy mortalities, low survival rates, and deformed larvae may be caused by the following:

1) Consistent larval survival below 50% suggests a water quality problem. Observations and measurements should be recorded for future reference.

2) Unusual shaped larvae is often a sign of contamination or disease. Presence of ciliates or other contaminating organisms indicates that culture water needs to be filtered.

3) Presence of a moderate number of trochophores after 24 hours indicates delayed development which can be caused by poor water quality, lighting, food supply and/or low temperature.

4) If there is an unusually large number of trochophores after 24 hours or a heavy mortality at 24 hours this can indicate poor quality eggs at fertilization. Keep in mind that trochophores do not feed and survive solely on energy reserves provided in the eggs. If the eggs are deficient, proper development will be hampered.

5) Early, rapid mortality may indicate contaminated or toxic conditions. It could be a toxic tank, plumbing, or residual chlorine. Always use nontoxic materials and make sure they are washed and aged with freshwater for 1-2 weeks before use. Repeated flushes may be needed.

6) Healthy larvae swim actively and have dark gut coloration. They often congregate on the surface with others trailing below the surface resembling a tornado. This is a healthy sign and referred to as "rafting."

7) The presence of numerous ciliates is a strong indication of bacterial contamination and warrants paying more attention to water exchanges and cleaning.

Chapter 11 - AMPHIPOD CULTURE

Introduction

Amphipods are highly diverse crustaceans under the order Amphipoda which contains nearly 7,000 species. There are four suborders including Gammaridea, Caprellidea, Hyperiidea and Ingofiellidea. Gammaridea is the largest and most diverse order with 5,500 described species. Environmentally they are found in all kinds of habitats including warm and cold terrestrial, marine and freshwater, shallow and deep environments. The majority live on seafloors in all latitudes although, there are also planktonic species. Terrestrial species live in moist environments on beaches, damp floors in fish hatcheries and under leaf litter in woods or under pots. Sizes normally range from 1-10mm (1/2") but deep sea species as large as 30 cm (15") have been captured (Museum 1996).

Basically they are detritivores ingesting a variety of organic particles or bacteria. Those living on marine and freshwater macroalgae (seaweeds) may be basically herbivores where as those on a mud or sand bottom feed on bacteria attached to surface areas. Others burrow 1-2 cm into the surface sediment to eat organic particles including bacteria while others are scavengers on dead animals and plants. *Gammarus locusta*, a marine specie, is often found under rocks and rubble at low tide. In the same area another specie *Orchestia sp.* reside in damp macroalgae mats along shorelines or attached to algae underwater. They feed on decaying macroalgae which is the major food in their diet yet, they will feed on carrion. They are negatively phototrophic and feed only at night. *Gammarus fasciatus*, a freshwater specie, can be collected from vegetation and leaf litter on the bottom of a pond or slow moving stream. *Leptochirus sp.* amphipods construct mucus tubes dwellings in mud bottoms of estuaries mud flats, salt creeks and marshes.

The most commonly cultured amphipod is the freshwater species *Hyalella azteca*. The family Hyalelidae has 35 species, all endemic to South America (mostly Lake Titicaca), except for *H. azteca* which found only in North America from Mexico to Alaska in virtually all permanent freshwaters (Blousfield, 1996). Densities of more than 10,000/m^2 have been observed in good habitats where they are important food for many birds, fish, salamanders, and large invertebrates. The closely related genus *Parahyalella* is found throughout the Caribbean in coastal waters and is usually found in high current areas clinging to macroalgae like *Sargassum* and *Gracillaria*.

Hyalella azteca has become widely used for testing the toxicity of sediments (USEPA, 1991). As result, starter cultures can be obtained from many government and commercial laboratories supplying

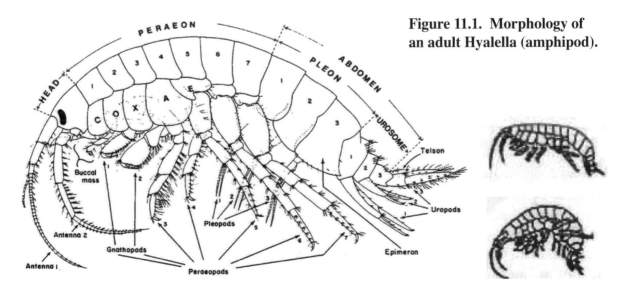

Figure 11.1. Morphology of an adult Hyalella (amphipod).

test organisms. Because *H. azteca* is common, it can also be collected from natural populations, but the taxonomic identity should be confirmed before putting extensive effort into culturing. In an animal with such a broad distribution, there is substantial variation among geographical strains in their optimum culture conditions. It would be prudent to test several strains in your culture system before choosing one to mass culture.

Recent interest in amphipod culture has come from the marine hobby trade. Seahorses readily accept amphipods as a food source and are a key live food item for survival success of adults and juveniles. As cautioned in the above paragraph as well as realizing there are over 7,000 species it is important to observe and understand the needs of the specie you are going to culture.

Life Cycle

Amphipods reproduce sexually with dimorphic males and females. Adult *Hyalella azteca* males (8 mm) are larger than females (6 mm) (Figure 11.1). Mature males of some species develop bulging eyes and chemical receptors on their antennae (Museum,1996). At 25°C, *Hyalella azteca* females reach sexual maturity in 28-33 days (de March, 1981). Females overwinter, producing their first broods in the spring when water temperatures consistently exceed 10°C. They reproduce continuously throughout the summer producing larvae that go through 5-6 instars to reach maturity. First instar stage is less then 1mm reaching a mature adult size of about 1.2cm in 7-8 weeks. Females that commence spawning in the spring die by the autumn, but females from late summer brood overwinter to begin the life cycle again the following spring. *Leptochirus* amphipods are said to reproduce a new generation every 30-40 days.

During reproduction (amplexus), the male grasps the female between its legs and swim around together. Males of some species seize immature females until their terminal molt when copulation occurs. Sperm is deposited into the female genital pore and eggs are fertilized as they pass through duct. The eggs are not attached as in crabs but are free floating within a cage-like brood chamber.

The fertilized eggs hatch, larvae grow and undergo metamorphosis within the brood pouch of females before they are released as true juveniles. One to 60 juvenile offspring are released in each brood at molting. Juveniles at times undergo parental care by some species.

Physical and Chemical Requirements

Hyalella azteca prefers temperatures exceeding 10°C and for optimum growth 20-26°C. If new populations need to acclimated to culture temperatures, acclimation should not exceed 2°C per day. In freshwater, *H. azteca* prefers hard (100-200mg/L) or alkaline (50-70 mg/L) conditions with slightly acidic pH (6.8). *H. azteca* commonly occurs in lakes and estuaries at 2-3 ppt salinity. It sometimes co-occurs with brine shrimp. It can be cultured in natural or reconstituted seawater up to about 15 ppt and can tolerate up to 30 ppt if acclimated slowly. Different strains probably prefer freshwater or estuarine conditions, so it is best to sample several geographical strains before choosing one to culture. *H. azteca* do best with 500-1000 lux full spectrum fluorescent light provided on a 16L:8D photoperiod. Dissolved oxygen should be >80% saturation which can be maintained with gentle aeration.

Culture Methods

We have pooled culture comments from popular aquarist literature as well as governmental sources. Amphipods live in a multitude of niche habitats and have different reproductive strategies, Determination of the specie you are working with is hard to determine unless you are a taxonomist. If collected by you from the wild in a specific situation you can made careful observations about their life-styles. The second option is to obtain stock from a source that maintains cultures and utilize the method they suggest. If in doubt, take a more holistic approach utilizing a little from each described culture method and make close observations.

Hyalella azteca, a freshwater specie, has been cultured in a variety of batch, flow-through, and recirculating systems (Environment Canada, 1997). Recommended water renewal rates are one tank volume per day, with a minimum of 30% per week. Since they are benthic, substrate is very important to amphipods. For *H. azteca*, presoaked and rinsed leaves (maple, birch, alder, poplar) have proven successful. Also acceptable are shredded paper towels, medicinal gauze, 210 μm mesh Nitex strips (2.5X2.5 cm), and plastic webbing. Perhaps artificial media used in biofilters like Bioballs also would be suitable. Harvesting is accomplished by shaking amphipods off the substrate, so the substrate must not be fixed to the bottom.

Suggestions for the culture of the marine specie *Gammarus locusta*, consists of a standard aquaria with a thin layer of coral rubble on the bottom. A lush growth of green macroalgae should be encouraged so position the tank in a direct sunlight for a portion of the day. Provide a large pore sponge filter and moderate aeration. Habitat consisting of loose or attached macroalgae is encouraged and algae encrusted media such as rocks. Feed small amounts of a variety of feeds including Spirulina powder or flakes, small amounts of dry fish food and raw chopped spinach. Capture by siphoning around the habitat and passing the water through a 300μ to 500μ mesh plankton collector or net. Often they reside in the course mesh of the sponge which can be squeezed to release them.

Amphipods of the family Talitridae and genus *Orchestia sp*. This specie resides in damp macroalgae mats along shorelines or attached to algae underwater. They feed on decaying macroalgae which is the major food in their diet yet, they will feed on carrion. They are negatively phototrophic and feed only at night. Sexual reproduction occurs within the damp algae mats or nests they construct on the sides of tanks along the surface. Females store fertilized eggs in the brood chamber and a clutch of 50 to 75 eggs is typical. Development occurs in the chamber and small juvenile amphipods mature in a matter of weeks under warmer temperatures. To culture provide a small separate container with a layer of damp, but not wet, decaying vegetative matter. Place the container within the tank and add water around the container. Within the water provide macro algae. If needed, supplement feeding within the container with algae based flake or frozen algae based foods.

Leptochirus amphipods reside on the bottom buried in the mud in mucus-like tubes. To setup a culture, collect or purchase fine silt mud from an estuary and sieve through a 500μ screen to remove larger particles, rocks, wood etc. To help reduce possible contamination place the liquid mud into a microwave and cook for 7 minutes or it can be placed in a freezer. Frozen mud may clump when defrosted and consistency can be restored by placing it in a blender for a minute or so. Place a thin (1/2") layer on the bottom of the aquaria or pan. When filling the tank with water fill slowly to prevent excessive turbulence. Allow micro and macroalgae to grow on three of the sides since this serves as a biofilter. A small sponge filter can also be added and turned off when adding food. Food can consist of routine additions of microalgae in conjunction or without fine ground flake foods containing vegetable matter. Allow either food to settle to the bottom. Care must be taken not to overfeed which is very easy to do. It may take about two months for a culture initially inoculated with 100-200 individuals to reach a harvestable stage. Amphipod density can be estimated by observing the amount of burrow holes on the surface of the mud. Harvest using a small siphon hose like airline tubing or a baster to suck the amphipod out of holes. Separate mud from amphipods using netting and feed to fish etc. Several cultures are suggested so alternating harvesting can be utilized. Water exchanges, 30% plus, should be made once or twice a month or when needed depending of water quality. Harvested amphipods, will often survive for 2 to 4 weeks in a refrigerator.

Usually juveniles and adults amphipods are both retained on a 275 μm mesh screen while adults can be harvested using 425 μm mesh screen or net. When multiple tanks are used each mature culture can be harvested approximately once per week. The largest culture tanks reportedly used are 50-80 L aquaria with amphipod densities of 10-50/L. It is not known whether such culture systems can be easily upscaled to mass cultures so this represents an area for further research.

Food Requirements

The key in feeding Amphipods is remember they are basically a benthic detritivore. Picking off pieces of organic particles of various living and nonliving surfaces including the glass sides of the culture vessel. *Hyalella azteca* derives most of its nutrition from algae and bacteria that adhere to organic particles <65 μm in diameter. In managed cultures, amphipods are fed a variety of diets including ground fish food flakes (20 mg/L, 1-3 times per week), dried and ground up maple, birch, alder, or poplar leaves, ground rabbit pellets, ground cereal leaves, ground fish food pellets, brine shrimp, heat-killed Daphnia, green algae and spinach, yeast and trout chow, *Spirulina* and fish food flakes (50-150 mg/L), diatoms (*Synedra*), and macro filamentous green algae like *Entromorpha* (USEPA 1991). Others suggest using the macro algae Ulva as suitable surface for feeding. Which diet is best will be determined by your specie used, size of culture system and the availability and cost to the components. Better water quality usually is maintained when additions of live micro or macroalgae are present in the culture tanks.

Amphipods as Live Feed in Aquaculture

Amphipods are a nutritious live food with a protein level of about 45%, fat 0.51%, carbohydrates 0.50%, ash 3%, fiber 0.53% and chitin 7.3%. However, Amphipods have not been used extensively in aquaculture and there are no reports of intensive mass culture. All of the culture methods described above were developed for the small scale production of amphipods for toxicity testing of feeding of special fish such as seahorses. There is therefore considerable room for improving these techniques for mass production which is more suitable for aquaculture. Recent rearing success of Seahorses has fueled interest in obtaining a continuous supply of Amphipods which is one of their favorite feeds.

Mathias and Papst (1981) stocked the amphipod *Gammarus lacustris* into ponds to produce forage for juvenile fish. Good, et al. (1982) compared several diets for the lobster *Homarus americanus*, one of which was the amphipod *G. oceanicus*. There was no significant difference in growth rate or mean weights on any of the diets, but the amphipod diet produced more darkly pigmented exoskeletons. Danielssen, et al. (1990) studied the feeding of turbot larvae (*Scophthalmus maximus*) in 2000 m³ outdoor mesocosms. Fish larvae began feeding on copepod nauplii, then switched to adult copepods, and finally to amphipods from day 20 onwards. Since amphipods are a natural food of many fish larvae, they could be useful as a final diet for fish to be released to the wild in restocking programs.

Tubicolous amphipod, *Cerapus* lives in a secreted
tube which it carries around. (Schmitt 1910)

Chapter 12 - MYSID CULTURE

Introduction

In their natural environments, many fish eat small, shrimp-like crustaceans called mysids. These estuarine and freshwater animals are not really shrimp, but their morphology is sufficiently similar that they are often referred to as mysid shrimp. There are more than 450 species of mysids, but only a few have been cultured. These include the marine and estuarine, species, *Mysidopsis bahia, M. almyra*, and *M. bigelowi* which are distributed along the coast of the Atlantic and Gulf of Mexico (Price, 1982; Price, et al. 1986). *Mysidopsis bahia* and *M. almyra* are commonly reared and used as indicator species in toxicity tests (Miller 1990). At many sites they co-occur and are so similar in form that accurate classification requires microscopic examination. A useful taxonomic key can be found in Price, et al. (1994). Mysids are also common in coastal waters of the Pacific, the species *Holmesimysis sculpta* and *Neomysis mercidis* being common in California. Mysids also occur in freshwater where the opossum shrimp *Mysis relicta* is common in North America and is a voracious predator on zooplankton.

Mysids are commonly collected from natural populations or they can be purchased from commercial aquaculture suppliers. Since *Mysidopsis bahia* is commonly used for screening toxicity in estuarine and marine waters, several commercial suppliers provide juveniles to testing labs (USEPA, 1993). However, these are usually expensive and not attractive for use as a food item because of cost.

Life Cycle

Mysids reproduce sexually with morphologically distinct males and females. The most practical characteristic used to identify females is the presence of eggs in the oviduct or egg sac. This bulging marsupium pouch (Figure 12.1) houses fry until they are fairly well developed juveniles thus the name opossum shrimp. Males are generally smaller than females and distinguished by their absence of a white brood pouch. *Mysidopsis* females mature at age 12-20 days at about 4 mm in length. Mating is at night and no eggs develop unless they are fertilized. The marsupium egg pouch is fully formed in females at about 15 days when they are 5 mm long. Females grow to about 9 mm in length and produce broods of upto 25-30 young at 17-20 day intervals (Ward, 1993).

Juveniles are planktonic for 1-2 days and then settle to the bottom and will feed on *Artemia* nauplii which are about 4-5 times smaller then juvenile mysids. In general, mysidopsis species are omnivorous and cannibalistic, feeding on diatoms and small crustaceans such as copepods (Mauchline 1980). While small and slow swimming, juvenile mysids are susceptible to cannibalism by adults. Because of the predatory behavior of adult mysids, they are useful as a live feed in aquaculture only when fish fry are big enough to consume prey at least 9 mm in length. Otherwise fry may become prey for the mysids.

Physical and Chemical Requirements

Mysids are tough not as forgiving as *Artemia* with regards to water quality. Optimum water temperature for mysids is 24-27°C (75-81°F) (Ward, 1982; 1993). They can acclimate to lower and higher temperatures, but temperature change should not exceed 3°C per 12 hours. Optimum salinity is 20-30 ppt and salinity change should not fluctuate above or below 3 ppt over a 12 hours period (Ward, 1984; 1991). Dissolved Oxygen should always exceed 5 mg/L.

Mysids do not like high light intensities which should range from 150-1000 lux on a 16:8 light/dark cycle (note normal sunlight is 10,000+ lux). Hemdal (2000) stated that 75 footcandles best. It is beneficial to have a 15-30 minute transition or maintain a constant low level, indirect, 15W "moonlight" when lights go on or off. A photoperiod of 14-16 hours of light is suggested.

The pH of seawater should be 7.8-8.2 and alkalinity should be less than 120 mg/L. Higher Nitrogen concentration should be closely monitored and maintained so that total ammonia and nitrite are less than 0.05 mg/L and nitrate is less than 18 mg/L. According to Hemdal (2000) they can be cultured in ammonia levels as high as 0.1 mg/l (1ppm) with no detectable adverse effects.

Separate and combined effects of changes in salinity and water temperature on the survival of laboratory hatched juvenile *Mesopodsis orientalis* were investigated. Full strength seawater (35/mil) was not favorable to juvenile survival. Salinities down to 10% seawater were tolerated when subjected to sudden exposure, but salinity acclimation increased juvenile ability to tolerate even fresh water. Water temperatures tolerated by the animals ranged from 12°C to 33°C. Salinities of 30% to 60% seawater and water temperatures of 22°C to 28°C were most favorable to the juveniles. Seawater of reduced salinity was found to be a major factor for occurrence of juvenile *M. orientalis* in abundance (Bhattacharya 2002).

Nutritional Value

Nutritional Analysis (Dry Weight Basis)		
Protein	78%	g/100g
Fat	21%	g/100g
Ash	5%	g/100g
Carbohydrate	<	g/100g
Energy	501	Cal/100g
Energy	2100	kJoules/100g

Table 12.1. Nutritional Analysis (dry weight basis) of freshwater Mysid shrimp (*Mysis relicta*). Collected from lakes in British Columbia, Canada.

Fatty Acid (Dry Matter Basis)	Name	mg/g	% Total
C14:0	Myristate	8	7.16
C16:0	Palminate	25	20.9
C16:1	Palmitoleate	12	9.8
C18:0	Stearate	1	0.58
C18:1	Oleate	17	14.4
C18:2 (omega 6)	Linoleate	7	6.24
C18:3 (omega 3)	Linolenate	7	6.03
C20:0	Arachidate	0.00*	0.00*
C20:1	Eicosaenoate	0.00*	0.00*
C20:3 (omega 3)	Eicosatrienoate	10	8.34
C20:4 (omega 6)	Eicosatetraenoate	0.00*	0.00*
C22:5 (omega 3)	Eicosapentaenoic	18	15.1
C20:5 (omega 6)	Docosapentaenoic	0.00*	0.00*
C22:5 (omega 3)	Docosapentaenoic	0.00*	0.00*
C22:6 (omega 3)	Docosahexaenoic	13	10.6
C24:0	Nervonate	0.00*	0.00*
			* = Not Detected

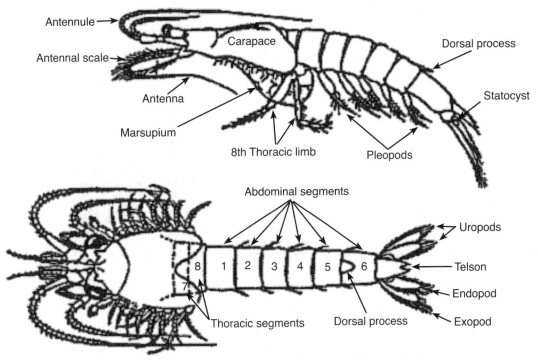

Antennule

Antennal scale

Antenna

Marsupium

Carapace

8th Thoracic limb

Dorsal process

Statocyst

Pleopods

Abdominal segments

Uropods

8 1 2 3 4 5 6

Telson

7

Endopod

Thoracic segments

Dorsal process

Exopod

Figure 12.1. Top and latertal view of mysid morphology

Culture Systems and Methods

Mysids have been cultured in natural and artificial seawater in flow-through and recirculating systems. Natural seawater should be 15 μm filtered to remove parasites and predators. Most mysid culture has been done to produce test animals to use in ecotoxicology. Consequently, there is little experience with mass cultures on the scale commonly required in aquaculture. Nonetheless, it is instructive to examine these small scale systems to get some ideas about how upscaling to mass cultures might be accomplished.

The largest of the small scale cultures has been done in 200 L (53 G) aquaria (USEPA, 1993). These are set up with two undergravel filters set at each end a bare spot in the middle for ease of capture of mysids. Undergravel filters are covered with a dolomite substrate and gentle aeration (Figure 12.2) is provided. Flow of at least 100 ml (3.5 fl. oz)/minute is introduced because mysids align with current for feeding. Three papers may be useful in designing a mysid culture system to fit your specific needs. Leger and Sorgeloos (1982) describe a batch culture method with reconstituted seawater, feeding by pumping *Artemia* nauplii from a refrigerator, and a separator for harvesting juveniles. Reitsema and Neff (1980) describe a recirculating system based on reconstituted seawater for *Mysidopsis almyra* culture. Their system consisted of a 57 L (15 G) aquarium used to maintain adult mysids joined to a 19 L (5 G) where mysid postlarvae were collected. A third recirculating seawater system using reconstituted seawater similar to the first two was described by Ward (1984).

Mysids are sensitive to handling so this should be kept to a minimum. A total harvest of a culture is achieved by using a 350 μm mesh nets. In a continuous culture the idea is to maintain adult brood stock and harvest new juveniles. Since they are sensitive to handling, normal netting and sorting can be detrimental to continuous reproduction. Leger and Sorgeloos, (1982) suggest a 1200 μm screen to allow juveniles to pass and retain adults.

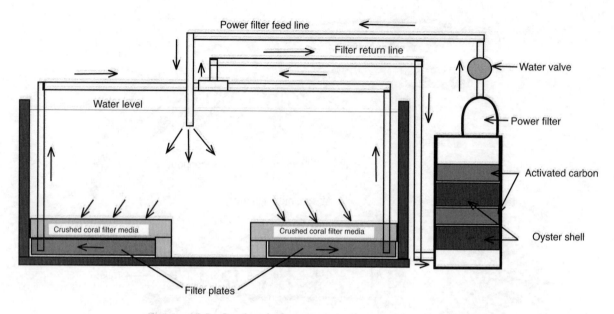

Figure 12.2. Recirculating system for mysid culture.

Cannibalism is a problem with mysid. Experiments conducted on other cannibalistic shrimp such a freshwater macrobrachium, has show that keeping light intensities low, adding habitat and good feeding decreases this tendency. We suggest using artificial or live macroalgae over the entire bottom. This will actually contribute in two ways. Increasing the vertical surface provides more active mechanical and biological filtration Secondly the vertical substrate provides a full water column full of hiding areas. Artificial grass can be made with plastic ribbons attached to egg crate light louver material. This arrangement can be easily lifted from the tank to allow for harvesting. A more natural habitat is *Caulerpa prolifera* is a good vigorous macroalgae that may also be considered. Commercial fish breeders have a problem with spawning pairs consuming newly spawned eggs or young fry. Often they use long bristle brushes (2-3") mounted on a stainless steel mounting wire in the center. This can be easily bent into the shape of the culture vessel. Like the artificial grass media it can be easily removed for harvesting and cleaning.

In addition to vertical substrate a thin layer 5 mm (3/16") of active coral hash sand rather then the typical thicker layer 2.5 cm (1") of dolomite or calcareous rocks is more active, easier to clean and control. Further filtration can achieved by adding small sponge filters suspended above the bottom. Sponge filters are ideal since they can be easily removed and cleaned. When cleaning sponge filters squeeze them in freshwater to remove particles and while still wet place them back into the tank. Do not let them dry out first since this will destroy the remaining live beneficial bacteria.

Since cannibalism of juveniles is a constant problem it is best to separate spawning adults from new hatch fry and juveniles. In laboratories egg-bearing females are often transferred into culture dishes just prior to spawning. Released postlarvae are then pipetted out and transferred into a grow-out tank. A more automatic and practical method (Figure 12.3) is to set two tanks at the same level side by side. Construct an air lift using 3/4" PVC pipe. Drill holes into submerged portion of the siphon and cover with 700-900μ screen (plastic window screen may work). Drill a hole at the top and put in a rigid air tube. Flow rates are regulated by how far the tube is submerged and how much air is provided. Do

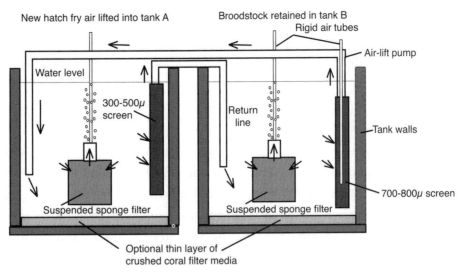

Figure 12.3. Airlift recirculating separating culture system for mysid shrimp culture.

not use an airstone. Small fry will be sucked into the tube and transferred into tank A. Put in a 3/4" to 1" return siphon between the tanks that is covered with a 300 to 500μ screen. Filtration and habitat can be added such as sponge filters and a thin layer of calcite/carbonate material.

Maximum density achievable with current strains is about 20 mysid/L. Mysid density can affect reproduction resulting in a higher proportion of females with empty brood sacs (Lussier et al. 1988). Optimal density in a open-flow system was about 15 adults/L and 10 adults/L in a static or recirculating culture system. Keeping adult densities relatively low is essential for constant production. Removal of new hatch postlarvae is essential to prevent cannibalism and overpopulation which will reduce reproduction. Siphoning one or two day old postlarvae are the easiest to remove since they often are attracted to the walls of the tank. Mysids that are already three to four days old will distribute into the water column, are more agile, thus harder to siphon. Sorting for smaller sizes using two nets together: one retains adult sizes and the other retains small sizes. A series of sieves or plankton collectors can also be used. Mysid populations need to be culled at least every two weeks to remove aging females and maintain vigorous reproduction. Probably the easiest method is to capture the entire population in the broodstock tank with a fine net and then pour them into larger mesh netting mounted on a clear tank of water. The larger mysids will remain in this net and the smaller will swim through the mesh. Larger individuals are then returned back to the broodstock tank and smaller ones can be easily captured and fed. This process can be repeated every 3-4 days without a significant detrimental effect.

To help keep reproduction steady, 20% of the population should consist of young mysids. From time to time reproduction may diminish. Several alternatives may help restore production. Culling males or those adults who do not show the typical female white brood pouch from the tank will reduce juvenile predation and competition. Ratios as low as 10:1 (females/males) will continue producing. Breaking down the tank and restarting with the right proportion of adults and juveniles may help. Water exchanges as high as 50% or more may help.

Harvestable cultures take time so mature. Allow 6-8 weeks before you need shrimp. When active broodstock number about 400-500 adults you should be able to harvest about 200 juveniles per day providing water quality and feeding remain good.

Culture tanks need to be free of hydroids and worms which are significant predators on mysids. These predators are more common in tanks that receive raw natural seawater or water from a tank with live rocks etc.

Food Requirements

Mysids have been traditionally cultured on *Artemia* nauplii which are fed at least once or twice per day. *Artemia* are stocked at 2-3/mL and a single mysid will consume about 150 per day. *Artemia* nauplii should be present at all times to help reduce cannibalism. Mysids also consume diatoms, like *Chaetoceros* or *Skeletonema*, which can be supplemented into their diets. Rotifers, *Brachionus plicatilis*, cultured on microalgae like *Nannochloropsis* and/or commercial feeds such as Roti-Rich™ or Culture SELCO™ can be used. Work still needs to be done on the use of microalgae and rotifers added as live feeds as to determine how much live *Artemia* can be replaced. Dry, prepared, commercial shrimp feeds is another option for consideration. Mysids, like shrimp, are browsers of surfaces and pick off particles. They do not normally consume food as a whole item but while browsing they gather and pack the particles of food into "balls" of food an then pull them apart and funnel them to their mouth. Or they will tear a food item apart. Therefor feed particles can be relatively small. Uneaten food and dead mysids should be removed from the culture tanks daily to minimize water quality problems.

Many references suggest enrichment of *Artemia* nauplii however they fail to caution against feeding and the affect on water quality. *Artemia* nauplii do not initiate feeding until they are about 12 hours therefor enrichment with oil based foods only coats the surface of the nauplii which washes off in the tank if they are not immediately consumed. Obviously this contributes to deterioration of water quality. Popular literature suggest a constant level of *Artemia* per ml however, pulsing or multiple smaller feedings are more efficient which regards to food quality, growth and water quality. If rotifers are used you must consider their high metabolic rate that must be addressed. The nutritional value diminishes rather quickly if not continuously fed. If you want to maintain a constant level of live food items then daily additions of microalgae should be used. It is suggested to refer to the rotifer and *Artemia* sections of this manual.

Mysids as Live Feed in Aquaculture

Mysid shrimp are highly nutritious. Fresh samples of *Mysis relicta* indicated 69.5% protein, 8.35% fat, 2.75% fibre and a maximum ash content of 5.5%. Mysid shrimp are currently being sold in the pet markets, both live and frozen. They are readily accepted by adult seahorses, and many other tropical reef fish. There are only a few examples of mysids being used as live feed in aquaculture. Kuhlmann, et al. (1981) added live mysids to the diet of juvenile turbot (*S. maximus*) and observed increased growth and food conversion rates. Reddy and Shakurtala (1986) used frozen mysids as food for juvenile *Penaeus merguiensis*. They reported that 1 kg of *P. merguiensis* consumed 6.5 kg mysids to produce 4.4 kg of new flesh for a conversion ratio of 1.5:1. However they provided no comparative data by which these results can be judged. Wickins, et al. (1995) found that a diet of mysids and *Artemia* produced the best growth in the larval culture of the lobster *Hommarus gammarus*. Seikai, et al. (1997) tested mysids and formula feed for rearing juveniles of the flounder *Paralicthys olivaceus*.

They compared four different sized fish reared on mysids or formula feed. Fish reared on mysids always had higher growth rates and better feeding efficiency than formula feed. The chemical composition and fatty acid profile of fish at the end of the experiment reflected their diet. Fish fed mysids had nearly the same composition as wild fish.

Furuta, et al. (1997) reported producing the flounder *Paralicthys olivaceus* in hatcheries for restocking natural populations. They observed low survival of flounder juveniles after release due to their poorly developed feeding ability on wild prey. Mysids are the natural prey of flounder juveniles, but they are only seasonally abundant. Consequently, when the mysid capture rate by juvenile flounders fell, the flounders became more vulnerable to predation by other fish. This illustrates a potentially important use of mysids in aquaculture. Stocking of hatchery reared juveniles into natural populations may be increased by introducing mysids into their diets just before release. Acclimation to this natural food item may substantially improve the efficiency of juvenile fish as predators in the wild.

In marine ornamental culture mysid are used to culture several unique species, including leafy sea dragons and cephalopods (Hanlon,Turk & Lee 1991). Common seahorses also accept juvenile and mysid adults as food. Basically most carnivorous adult size fish will accept mysids as food.

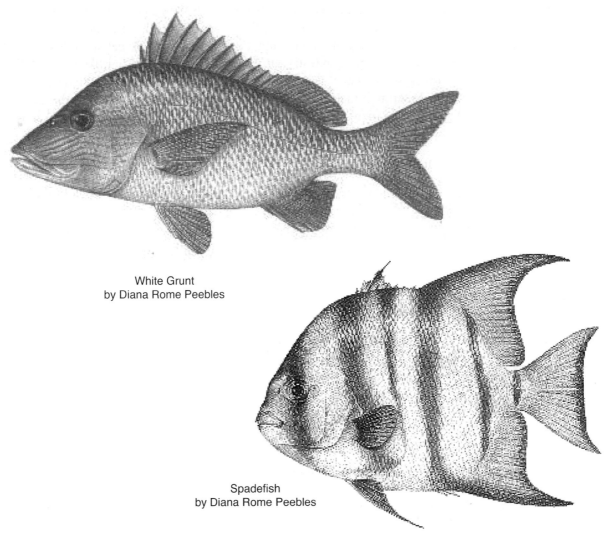

White Grunt
by Diana Rome Peebles

Spadefish
by Diana Rome Peebles

Chapter 13 - MICRO WORMS

Introduction

Micro worms, *Anguillula silusiae,* or *Panagrellus redivivus*, are tiny cylindrical, milk-white colored, nematodes that are distantly related to seat-worms (*Tricbina*) or pork-worms (*Ascaris*). They are closely related to the 2 mm long vinegar-eel (*Anguillula aceti*). They differ only in minor characteristics (Campbell, 1949). Often they are referred to as thread worms, micro-eels or vinegar-eels. Nematodes are universal and found in nearly every moist soil sample or natural water sources. Nearly every plant and animal has it's nematode inhabitants. Micro worms are so small that many fish fry will take them without having to feed on infusorial feeds. They range in size from 1.5 mm to 3 mm (1/16" - 1/10") depending on species. Although they are a freshwater animal they may also have applications in salt-water culture.

They move constantly and would be attractive to many prey species. Campbell (1949) noted that micro worms are used as food for newly-hatched dwarf Cichlids (*Apistogramma ramirezi*), Flame Tetras (*Hyphessobrycon flammeus*) and some of the Barbs. He felt that they would be excellent feed for small mouthed adult fish such as Neons, Pencil Fishes, Glass Fish, Zebras, White Clouds and so-called *Kublii* Loach.,

Life Cycle

Ivleva (1969) provided excellent information about their life cycles which is summarized in the following text. Micro worms show pronounced sexual dimorphism (Figure 13.1). The males are smaller, more slender, and have a curved tail. The genitalia occupy the entire space between the intestine and body wall. Male genitalia comprise an unpaired testis and attachment setae (stiff hairlike or bristlelike structure) which continues into the seminal duct which opens into the anal duct. Female unpaired ovary and their wide uterus are visible through the body walls. Ovaries and the uterus are usually packed with eggs and embryos. Proportion of sexes are unequal, usually 3.75 females per male. As the culture matures this ratio drops to 1.2 females per male but never exceed the female population.

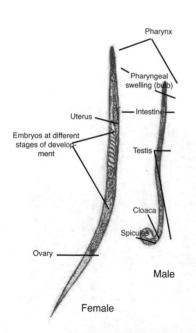

Fig 13.1 Male and female micro worms P. redivivus (modified from Ivleva 1969)

Eggs develop in the large uterus and batches of fertilized eggs enter the lower uterus daily. Egg cleavage begins in the upper portion of the uterus and gradually move to the middle and posterior parts during development.

Complete development is in 2 to 2.5 days at 20°C (68°F). All stages of development from eggs to embryos is retained within the uterus. Young nematodes move freely and escape via the vagina.

The number of progeny per litter varies with body length and age of the female. Young nematodes (1.45 mm) produce 7 to 19 progeny while older, larger females produce up to 42 per litter. Progeny are produced every 1 to 1.5 days and a female produces about 15 litters in a lifetime resulting in about 300 progeny. *P. redivivus* molt about 4 times before reaching sexual maturity.

Newly hatch progeny are 180μm to 290μm long and 0.13μm round. At 20°C (68°F) their size increases by a factor of 3 by the end of the first day and 5-6 during the next 3-4 days. About 90% of the population attain sexual maturity in 3 days at a body length of 1.10 to 1.65 mm. Differential growth occurs between sexes starts early in development. Early males average length is 1.32 mm and range from 1.18 to 1.40 mm and diameter of 50μm. While females average 1.43 and range from 1.10 to 1.65 mm and diameter of 54μm. At day 6 this ratio changed from 1.44 mm for males and 1.61 mm for females. Maximum lengths are 2.05 mm for males, body diameter of 0.73μm and 2.76 mm and diameter of 0.95μm for females .

Life span can vary depending on temperature. At a lower temperature of 7.5°C (46°F) the population shifts to about 40% more older individuals than those grown at 20°C (68°F).

Physical and Chemical Requirements

The rate of growth depends on temperature and the presence of food and moisture. Studies on *P. redivivus* (Ivleva, 1969) showed that a stable mixed age culture with an initial density of 300-350/cubic cm increased by a factor of 400 in density and a 7-8 increase in biomass within 10 days at 20.5°C (70°F). Within the next 30 days, along with partial replacement of food substrate, the culture can achieve a density limit of 450,000-500,000/cubic cm and a biomass of 1.41 mg/cubic cm. A *P. redivivus* population can only survive for a brief period at this density and fecundity drops significantly. At 8°C (46°F) growth rate decreases to one-half at taking 42 days to reach a density of 360,000/cubic cm. At 1.5°C (35°F) the growth rate slows further taking 58-60 days to reach a density of 183,000/cubic cm. Popular aquarium literature supports even higher temperatures ranging to 30°C but warns against quick fouling and death of the culture.

We recommend keeping the temperature around 21° to 27°C (70° to 81°F) and prefer the lower to mid portion of this range. However, as shown growth rates can be controlled by lowering the temperature which is good when regulating production with regards to timing when microworms are needed.

Light period is not clearly specified however, direct sunlight is not recommended. Campbell (1949) suggest a dark area whereas, Piparo (1979) suggests a well lighted area using a fluorescent light but does not specify duration or intensity. Ivleva, 1969 demonstrated that *P. redivivus* is highly resistant to light and found that illumination exerts no influence whatsoever on population growth. We suggest moderate to low, non-sun, lighting for 10-12 hours daily. Lighting, located above the culture, can be just a simple cool-white fluorescent bulb.

Although micro worms live in decaying and fermenting substrates with masses of microorganisms with a high biological oxygen demand (BOD) they cannot live in anaerobic conditions. Masses of these worms crawl to surfaces and walls when the oxygen levels quickly deteriorate. Closely related vinegar eel worms (*Turbatrix - Anguillula aceti*) can live from 7 to 14 days in total absence of oxygen (Brand, 1946). However, we feel oxygen is important and low circulation in the culture container is suggested. Drilling small holes in the lid may help.

P. redivivus lives and reproduces in an acidic medium within 3.35 to 3.60 pH range. During transfer into fresh substrate they are subject to neutral (7.0) or slightly alkaline levels. However, in 3-4 days the medium again becomes acetic. Although this species tolerates a wide range of pH, optimal growth is achieved in 3.4-4.2 pH (Morton and Cook, 1966). Other nematode species show similar preference for low pH.

Ammonia, nitrites, nitrates etc. are not discussed and at first may appear to be not overly important. However, rapid fouling is a well documented problem resulting in a quick, mass mortalities.

Culture Systems

Simple is the word with regards to a culture vessel. Culture vessels are highly variable from modified sauces, petri dishes to plain plastic shoe boxes. Containers must be made of glass or plastic. All require loose to moderately tight covering to reduce excessive evaporation of the culture media yet allow for "breathing" or minimal air exchange. Most containers mentioned in the literature have a high surface area and moderate depth of 6 to 12 cm (2 3/8" x 4 3/4"). The preferred culture media depth is from 1 to 15 cm (3/8" to 6") so the culture vessel should have sides at least 3 to 6 cm (1 1/4" to 2 3/8") above the media.

As the culture matures micro worms will crawl off the media surface onto solid moist surfaces making it easy to harvest. Several methods of collection are mentioned one consists of plastic flat wood strips (3-4 cm width) place on top of the culture media with spaces in between. Others use high side wall containers and collect the worms off the sides. Collecting can be accomplished using Q-tips, a brush, feathers, wooden or rubber spatulas or your finger. Harvesting is accomplished by simply scraping and washing them into the tank or into a jar with water and then transferred into the rearing containers.

Food Requirements

Food is easily obtained in any grocery store. Products used are usually processed grain feeds. Most prefer Pablum™ (baby oatmeal) and one-part Brewer's yeast (activated yeast). Pablum produces a rich and more rapid growth media. Klee (1959) stated that prepared baby foods contain enough yeast therefore, may not be required. However, if just plain oat meal is used it will require the addition of activated yeast.

Piparo (1979) suggests the following procedure to fill a plastic shoe box. Pour the entire contents of yeast (1/4 oz. pack) into 16 oz. of warm tap water and stir until all is dissolved. Some suggest stale, slightly sour beer (Henzelmann, 1960) or milk (Ivleva, 1969) in lieu of water. To the 16 oz. of water, add 8 oz. of Pablum (2:1 ratio). Stir until all the meal is moistened. The mixture should have a very thick, paste-like consistency and emit a yeasty odor. A pencil should be able to stand erect in the media. Add more water or meal as needed. Consistency of the media is important. If too liquid, a bacteria film develops within 24 hours and the growth is dramatically slowed.

Spoon prepared mixture into culture container trying not to get any on the sides. Ivleva (1969) noted that micro worms mainly inhabit the top 5 mm (3/16") but the layer which they occupy during the period of intensive growth is only 2-3 mm. (1/16-1/8") suggesting a minimum layer of 1.0 to 1.5 cm. (7/16"-5/8"). Thinner layers dry out rapidly and require more frequent culture transfers. Clean sides will eliminate getting culture media in your fry tanks which may cause fouling. Spread mixture evenly and inoculate with a worm culture over the entire surface. Keep culture covered.

Culture Methods

Once a culture is setup it takes about a week to become very active. When the culture reaches maturity worms will begin to crawl up on the sides of the culture container or surface plates. This reaction is an attempt to escape from the medium when their density gets too great and oxygen levels are low. These are normally fed directly without any other cleaning suggested unless you scrape a significant amount of culture food media when collecting the worms.

Growth rates and life span depend mainly on temperature and the nature of the medium. For *P. redivivus*, culture densities rise rapidly to a peak within 10-11 days at 18-20°C (64-68° F). Under these conditions the rich mixture of microorganisms mineralize the medium so that the nutrient substrate liquefies, becomes dark brown to gray (depending on media used), and has an acrid, foul odor which is quiet different the good yeasty odor. At this stage the peak population exist for only a brief period then decrease drastically over a 24-25 day period (Ivleva, 1969). Under these conditions a culture cannot survive for more than 35-40 days.

At this point the upper surface of the culture will become crusty. However, before reaching this point a culture can be revived by addition of new media. This will lengthen the live span of a culture for around 3 months. Ivleva (1969) renews cultures by adding just dry oat flakes or oat meal (Pablum™) around the 20 to 25th day of culturing. This stimulates regrowth of the culture which attains a peak after 50-60 days at 18-20°C (64-68° F).

Piparo (1979) suggest that before the foul odor, considered the point of no return, a culture must be restarted. At the point when the media becomes runny this indicates the solid feed (Pablum™) and yeast are becoming exhausted. Simultaneously, the productions of worms begins to diminish. At this point you can extend the culture by pouring dry Pablum (baby oatmeal) into the culture and stir the entire culture until the new food is thoroughly moistened and is a thick paste. Using this technique the culture can be harvest in about a week and will last for about 6 weeks at 28°C (82°F).

Henzelmann (1960) describes a uniquely different culture method using stale and slightly sour beer. He adds just enough beer to dry oat meal so that the surface area will remain dry. Apparently there is no mixing like others and no yeast is added. Temperature he suggests is 18-22°C (64-74°F). Within 2 weeks the worms will cover the middle and sides of the container. Replenish the beer only enough to remoisten the lower layer but allowing the surface to remain dry. Using this method he has maintained the same culture for 1 1/2 years. He does not mention replenishment of the food media but is assumed to be necessary.

Campbell (1949) found that restarting a new culture was enhanced when 1-2 tablespoons of the liquid from old cultures is added. Sufficient worms and microorganisms will be present in this liquid inoculant to start good culture.

Storage of Inactive Cultures

Ivleva (1969) maintained cultures of *P. redivivus* without losses for 6 months or longer at near-zero temperatures. At these temperature and after 170 days of culturing, populations have a density of about 50,000 organisms per sq. meter. The population consisted of mainly sexually mature individuals (about 65%) and juveniles of near-adult size (about 35%). There were no newborn. Transfer to a normal environment restored rapid growth and masses of young appeared with 3-4 days. Cultures were ready for harvesting in 10-11 days.

Ivleva (1969) further pointed our that the simultaneous application of cold and drying is especially promising. Under these circumstances nematodes can be kept for as long as 1.5-2 years. However, constant air humidity is essential. Nematodes can be revived by a gradual warming and moistening of the substrate. The revived worms are juveniles aged 1-1.5 days at a density of 100-200 worms per square centimeter.

Campbell (1949) mentioned that these worms have been known to freeze solid without being killed. Others mention worms being obtained from a rich dry peat medium when moistened.

Food Value in Aquaculture

Little has been done for the *P. redivivus* discussed by Ivleva (1969). However, other species of cultured nematodes of the micro group comprising the genera *Rhabditis* and *Turbatrix* are related to *Panagrellus*. Knaack (1958) reported that the body consists of 76% water and 24% dry material which consisted 40% protein and 19.5% fat. They are rich in fat compared with other invertebrates and are a good initial food for young larvae which initially demand a higher fat level.

Henzelmann (1960) mentions that microworms are suited as great carriers of vitamins. Experiments in supplementing stale beer food formula with vitamins A, B, C and D resulted in higher quality larval fish with no defects.

As shown below, (Table 13.1) yields can be significant within a specific period of time without replenishment and at a temperature of 20°C (Ivleva 1969).

Days of Culture	Thousands of worms/cubic cm	Biomass, mg/cubic cm	Biomass, g/meter squared
0	0.34	0.3	1.5
6	22.5	21.3	108.0
11	139.5	133.9	669.5
25	74.0	71.0	355.0

Table 13.1 - Yields of microworms at 20°C (68°F) over a 25 day period without feed repleshment.

Chapter 14 - LITERATURE CITED

Achuthankutty, C.T., Y. Shrivastava, G.G. Mahambre, S.C. Goswami, and M. Madhupratap. 2000. Parthenogenetic reproduction of *Diaphanosoma celebensis* (Crustacea: Cladocera): influence of salinity on feeding, survival, growth and neonate production. Marine Biology 137:19-22.

A.C.T.E.D. 1997. Manual for farming the hard shell clam in Florida. Harbor Branch Oceanographic Institution Inc. Aquaculture division, Ft. Pierce, Florida.

Alam, M. J., K. J. Ang and S. H. Cheah 1993a. Use of *Moina micrura* (Kurz) as an *Artemia* substitute in the production of *Macrobrachium rosenbergii* (de Man) post-larvae. Aquaculture 109: 337-349.

Alam, M. J., K. J. Ang and S. H. Cheah 1993b. Weaning of *Macrobrachium rosenbergii* (de Man) larvae from *Artemia* to *Moina micrura* (Kurz). Aquaculture 112: 187-194.

Allen, G.R., 1972. The Anemonefishes. T.F.H. Publications, Inc. Ltd. New Jersery.

Alver, M.O., J.A. Alfredsen, and Y. Olsen. 2006. An individual-based population model for rotifer (Brachionus plicatilis) cultures. Hydrobiologia 560:93-108.

Angelos A. J.M. Lotz, J.T. Ogle and J.T. Lemus 2004. Effect of diet on the fucundity of *Acartia Tonsa*, a calanoid used for red snapper culture. To be presented to the World Aquaculture Society 2004 conference in Hawaii.

Aoki, S. and A. Hino 1996. Nitrogen flow in a chemostat culture of the rotifer *Brachionus plicatilis*. Fisheries Science 62: 8-14.

Aoki, S., J. Kanda and A. Hino 1995. Measurements of the nitrogen budget in the rotifer *Brachionus plicatilis* by using ^{15}N. Fisheries Science 61: 406-410.

Arimoro, F.O. 2006. Culture of the freshwater rotifer, Brachionus calyciflorus, and its application in fish larviculture technology. African Journal of Biotechnology 5:536-541.

Bainbridge, R., G.C. Evans, and O. Rackham. 1969. Light as an Ecological Factor, British Ecological Society, Symposium #6, Blackwell Scientific Publications, Oxford, London: 452p.

Barclay, W. and S. Zeller. 1996. Nutritional enhancement of n-3 and n-6 fatty acids in rotifers and *Artemia* nauplii by feeding spray-dried *Schizochytrium* sp. J. World Aquaculture Society 27:314-322.

Barnabe, G. 1990. Harvesting micro-algae. In: Aquaculture, Vol. 1. Barnabe, G. (Ed.), Ellis Horwood, New York, pp 207-212.

Bardach, J. F., J.H. Ryther and W.O. McLarneu. 1972. Aquaculture: The farming and husbandry of freshwater and marine organisms. John Wiley & Sons, New York, NY. Book 868 pgs.

Barnes, R. D. 1963. Invertebrate Zoology. W. B. Saunders Company, London. Book 632 pgs.

Belcher, H. and E. Swale, 1988. Culturing algae: a guide to schools and colleges. Pub. by Inst. of Terrestrial Ecology. Cambridge England.

Bellows, W.K. and R.C. Guillard, 1988. Microwave sterilization. J. Exp. Mar. Bio. Ecol. 117:279-283.

Ben-Amotz, A., R. Fishler and A. Schneller. 1987. Chemical composition of dietary species of marine unicellular algae and rotifers with emphasis on fatty acids. Marine Biology 95:31-36.

Berg, C.J., Jr. 1983. Culture of Marine Invertebrates, Selected Readings, Hutchinson Ross Publishing Company, Stroudsburg, Pennsylvania: 386p.

Berghahn, R. S. Euteneuer and E. Lubzens. 1988. High density storage of rotifers (*Brachionus plicatilis*) in cooled and undercooled water. European Aquaculture Society, Special Publication.

Berglund, D., Cooksey, B. K., Cooksey, E. and Priscu, L. R. 1987. Collection and Screening of Microalgae for Lipid Production. SERI/SP-231-3071. Solar Energy Research Institute, Golden, CO. 41-52.

Bhattacharya S.S. 1982. Salinity and Temperature Tolerance of juvenile *Mesopodopis orientalis* laboratory study. Hydrobiology Vol 93. 3 edition, August 1982, Publisher Springer Netherlands

Biedenbach, J. M., L. L. Smith, et al. 1990. Use of a new spray-dried algal product in penaeid larviculture. Aquaculture 86: 249-257.

Birky, C.W. Jr. and J.J. Gilbert 1971. Parthenogenesis in rotifers: The control of sexual and asexual reproduction. American Zoologist 11:245-266.

Blaxter, J.H.S. 1965. The feeding of herring larvae and their ecology in relation to feeding. Calif. Cooperative Oceanic Fisheries Investigations Reports 10:79-88.

Boehm, E.W.A., O. Gibson, and E. Lubzens. 2000. Characterization of satellite DNA sequences from commercially important marine rotifers *Brachionus rotundiformis* and *B. plicatilis*. Marine Biotechnology 2:38-48.

Bold, H. C. and Wynne, M. J. 1985. Introduction to the Algae: Structure and Reproduction. 2nd ed. Prentice-Hall, Inc., Englewood Cliffs, NJ. 722 pp.

Boney, A. D. 1983. The Institute of Biology's Studies in Biology no. 52 Phytoplankton. Photobooks (Bristol) Ltd. ISBN # 0-7131-2476-8.

Bonou, C.A., and L. Saint Jean. 1998. Regulation mechanisms and yield of brackish water populations of *Moina micrura* reared in tanks. Aquaculture 160:69-79.

Bosque, T., R. Hernandez, R. Perez, R. Todoli, and R. Oltra. 2001. Effects of salinity, temperature and food level on demographic characteristics of the seawater rotifer *Synchaeta littoralis* Rousselet. J. Exp. Mar. Biol. Ecol. 258:55-64.

Bourrelly, P. 1968. Les Algues d'eau douce. Initiation a la systematique. II: Les algues jaunes et brunes, Chrysophycees, Pheophycees, Xanthophycees et Diatomees. Ed. N. Boubee and Cie, Paris. 438 pp.

Brock, T. D., Smith, D. W. and Madigan, M. T. 1984. Biology of Microorganisms. Prentice-Hall, Inc., Englewood Cliffs, NJ.

Brand, T. 1946. Anaerobiosis in Invertebrates. biodynamica, Normandy.

Brown, M. R. 1991. The amino acid and sugar compositions of 16 species of microalgae used in mariculture. Aquaculture, 145: 79-99.

Brown, M. R. and S. W. Jeffrey 1992. Biochemical composition of microalgae from the green algal classes Chlorophyceae and Prasinophyceae. 1. Amino acids, sugars and pigments. J. Exp. Mar. Biol. Ecol. 161: 91-113.

Brown, M. R., S. W. Jeffrey, J. K. Volkman, and G. A. Dunstan. 1997. Nutritional properties of microalgae for mariculture. Aquaculture 151: 315-331.

Buskey, E. J., C. Coulter and S. Strom 1993. Locomotory patterns of microzooplankton: potential effects on food selectivity of larval fish. Bull. Marine Sci. 53: 29-43.

Butcher, R. W. 1967. An Introductory Account of the Smaller Algae of British Coastal Waters. Part IV: Cryptophyceae. Ministry of Agric., Fish., and Food, Fish. Invest., Ser. IV, 54 pp.

Camacho, E. M., M. E. Martinez, et al. 1990. Continuous culture of the marine microalga *Tetraselmis* sp. - productivity analysis. Aquaculture 90: 74-84.

Campbell, Arthur S. 1949. Micro Worms. The Aquarium Journal, San Francisco Aquarium Soc. Vol. XX#4, pp-99101.

Chandler, G.T. 1986. High-density culture of meobenthic harpacticoid copepods within a muddy sediment substrate. Cand. J. Fisheries Aquatic Sci. 43:53-59.

Chen, J.F. 1991. Commercial production of microalgae and rotifers in China. In: Rotifer and Microalgae Culture Systems. The Oceanic Inst., Hawaii. pp 105-112.

Cheng, S.H., T. Suzaki, and A. Hino. 1997. Lethality of the Heliozoan *Oxnerella maritima* on the rotifer *Brachionus rotundiformis*. Fisheries Sci. 63:543-546.

Cheng, S.H., S. Aoki, M. Maeda and A. Hino. 2004. Competition between the rotifer Brachionus rotundiformis and the ciliate *Euplotes vannus* fed on two different algae. Aquaculture 241:331-343.

Ciros-Pérez, J., Gómez, A., Serra, M., 2001. On the taxonomy of three sympatric sibling species of the *Brachionus plicatilis* (Rotifera) complex from Spain, with the description of *B. ibericus* n. sp. J. Plankton Res. 23, 1311-1328.

Corkett, C. J. 1967. Technique for rearing marine calanoid copepods in laboratory conditions. Nature, London. 216, 58-59.

Coughlin, D.J., J.R. Strickler, and B. Sanderson. 1992. Swimming and search behavior in clownfish, *Amphiprion perideraion*, larvae. Animal Behavior 44:427-440.

Coughlin, D.J. 1993. Prey location by clownfish *(Amphiprion perideraion)* larvae feeding on rotifers *(Brachionus plicatilis)*. J. Plankton Research 15:117-123.

Creswell, L., D. Vaughan, and L. Strurmer, 1990. Manual for the cultivation of the American oyster, Crassostrea virginica, in Florida. Aqua. Mark.. Develop. Aid Program, Fl Dept. Ag. and Consumer Ser., Tallahassee, FL.

Cotonnec G., C. Brunet, B. Sautour and G. Thoumelin, 2001. Nutritive Value and Selection of Food Particles by Copepods During a Spring Bloom of Phaeocystis sp. in the English Channel, as Determined by Pigment and Fatty Acid Analyses. Journal of Plankton Research Vol.23 no.7 pp.693-703, 2001

Coutteau, P. 1996. Micro-Algae. Manual on the production and use of live food for aquaculture. FAO Fisheries Technical Paper 361. pp 7-48. Lavens P. and P. Sorgelos editors.

Creswell, LeRoy, D. 2001. Aquaculture Desk Reference. Published by Florida Aqua Farms Inc. Edited by Frank Hoff. 206 pages.

Creswell. LeRoy, D. Vaughn, and L. Sturmer. 1990. Manual for the cultivation of the American oyster, *Crassostera virginica*, in Florida. Aquaculture Marketing Develop. Aid Progam. FL Dept. Ag. & Consumer Ser. Tallahassee, FL.

Dabrowski, K. and D. Culver 1991. The physiology of larval fish. Aquacult. Mag. 17: 49-61.

Danielssen, D., A.S. Haugen, and V. Oeiestad. 1990. Survival and growth of turbot (*Scophthalmus maximus* L.) in a land-situated mesocosm. Floedevigen Rapportserie 2:11-45.

Darley, M.W., 1982. Algal Biology: A Physiological Approach, Blackwell Scientific Publications Oxford, London:168 p.

Dawes, C. J. 1981. Marine Botany. Wiley and Sons. New York. 628 pp.Dewey, I.E. and B.L. Parker. 1964. Mass rearing of *Daphnia magna* for insecticide bioassay. Journal Economic Entomology 57(6).

de March, B.G.E. 1981. *Hyalella azteca*. pgs. 61-77, In: Manual for the Culture of Selected Freshwater Invertebrates. Canada Spec. Publ. Fish. Aquatic Sci., No. 54, Dept. Fisheries and Oceans, Ottawa, ON, Canada.

Dendrinos, P. and J. Thorpe., 1987. Experiments on the artificial regulation of the amino acid and fatty acid contents of food organisms to meet the assessed nutritional requirements of larval, post-larval and juvenile Dover sole, *Solea solea* (L.). Aquaculture, 61: 121-154.

Derry AM, Hebert PDN, Prepas EE (2003) Evolution of rotifers in saline and subsaline lakes: A molecular phylogenetic approach. Limnol Oceanogr 48:675-685

Dhert, P. 1996. Rotifers. Manual on the production and use of live food for aquaculture. FAO Fisheries Technical Paper 361. pp 49-77. Lavens P. and P. Sorgelos editors.

Dhert, P., R. B. Bombeo, P. Lavens, and P. Sorgeloos 1992. A simple semi flow-through culture technique for the controlled super-intensive production of *Artemia* juveniles and adults. Aquaculture Engineering 11: 107-119.

Doohan, M. 1973. An energy budget for adult *Brachionus plicatilis* Muller (Rotatoria). Oecologia (Berl.) 13:351-362.

Douillet, P. 1998. Disinfection of rotifer cysts leading to bacteria-free populations. J. Exp. Mar. Biol. Ecol. 224:183-192.

Doulliet, P. 1987. Effect of bacteria on the nutrition of the brine shrimp *Artemia* fed on dried diets. *Artemia* research and its applications. Vol. 3. Ecology, culturing, use in aquaculture. Wetteren, Belgium, Universa Press. 295-308.

Douillet, P.A. 2000. Bacterial additives that consistently enhance rotifer growth under synxenic culture conditions 2. Use of a single and multiple bacterial probiotics. Aquaculture 182:241-248.

Droop, M. R. 1974. Herterotrophy of carbon. In: W.D.P. Stewart (Ed). Algae Physiology and Biochemistry (Botanical Monographs Vol. 10). Univ. of Calif. Press, Los Angeles. pp 530-559.

Drouet, F. 1968. Revision of the classification of the Oscillatoriaceae. Monogr. Acad. Natural Sci. of Philadelphia, No. 15. 370 pp.

Drouet, F. 1973. Revision of the classification of the Nostocaceae with cylindrical trichomes. Hafner Press. New York. 292 pp.

Dunstan, G. A., J. K. Volkman, S. W. Jeffrey and S. M. Barrett 1992. Biochemical composition of microalgae from the green algal classes Chlorophyceae and Prasinophyceae. 2. Lipid classes and fatty acids. J. Exp. Mar. Biol. Ecol. 161: 115-134.

Dupree, H. K. and J. V. Hunter, 1984. Third Report To The Fish Farmers, US Dept. of the Interior, Fish and Wildlife Service, 270 p.

Dutrieu, J. 1960. Observations biochimiques et physiologigues sur le development d" *Artemia salina*. Arch. Zool. exp. gen., 99, 1-134.

Environment Canada 1997. Biological Test Method: Test for survival and growth in sediment using the freshwater amphipod *Hyalella azteca*. Report EPS 1/RM/33.

Escobal, P., 1993. Inside ultraviolet sterilizers. Aquarium Fish Magazine. Jan 1993. pp 52-63.

Eversole, A. G., Michener, W. K. and Eldridge, P. J. 1980. Reproduction cycle of *Mercenaria mercenaria* in a South Carolina estuary. Proc. Nat. Shellfish Assoc. Vol. 70, Pgs 22-30.

Fengqi, L. 1996. Production and Application of Rotifers in Aquaculture. Aquaculture Magazine 22(3):16-22.

Fernandez-Diaz, C., E. Pascual and M. Yufera 1994. Feeding behavior and prey size selection of gilthead seabream, *Sparua aurata*, larvae fed on inert and live food. Marine Biol. 118: 323-328.

Fernandez-Reiriz, M. J., U. Labarta and M. J. Ferreiro 1993. Effects of commercial enrichment diets on the nutritional value of the rotifer (*Brachionus plicatilis*). Aquaculture 112: 195-206.

Fogg, G.E. and Than-Tun. 1960. Interrelations of photosynthesis and assimilation of elementary nitrogen in blue-green algae. Proceedings Royal Society, B 153:111-127.

Fogg, G.E. 1959. Nitrogen nutrition and metabolic patterns in algae. Symposium Society Experimental Biology 13:106-135.

Fogg, G.E. 1963. The role of algae in organic production in aquatic environments. British Phycological Bulletin 2:195-205.

Fogg, G.E. 1966. Algae cultures and phytoplankton ecology. Univ. Wisconsin Press, Madison, WI, 126 p.

Fontaneto, D., Giordani, I., Melone, G., Serra, M., 2007. Disentangling the morphological status in two rotifer species of the *Brachionus plicatilis* species complex. Hydrobiologia 583, 297-307.

Fox, J.M. 1983. Intensive algal culture techniques. In: CRC Handbook of mariculture. Vol. 1. Crustacean Aquaculture. McVey J.P. (Ed.). CRC Press, Inc., Boca Raton, FL, pp 229-236.

Frolov, A. V., S. L. Pankov, K. N. Geradze and S. A. Pankova 1991. Influence of salinity on the biochemical composition of the rotifer *Brachionus plicatilis* (Muller). Aspects of adaptation. Comp. Biochem. Physiol. 99A: 541-550.

Fu, Y., K. Hirayama, et al. 1991a. Morphological differences between two types of the rotifer Brachionus plicatilis O.F. Muller. J. Exp. Mar. Biol. Ecol. 151: 29-41.

Fu, Y., K. Hirayama, et al. 1991b. Genetic divergence between S and L type strains of the rotifer *Brachionus plicatilis* O.F. Muller. J. Exp. Mar. Biol. Ecol. 151: 43-56.

Fujita, S. 1979. Culture of red sea bream, *Pagrus major*, and its food. In: Cultivation of Fish Fry and Its Live Food, E. Styczynska-Jurewicz, T. Backiel, E. Jaspers and G. Persoone (eds.), European Mariculture Society, Special Publication 4:183-197.

Fukusho, K. 1980. Mass production of a copepod, *Tigriopus japonicus*, in combination with the rotifer *Brachionus plicatilis*, fed yeast as a food source. Bull. Japan. Soc. Sci. Fish. 46: 625-629.

Fukusho, K., Arakawa, T. and Watanabe, T. 1980. Food value of a copepod, *Tigriopus japonicus*, cultured with *ω*-yeast for larvae and juveniles of mud dab, *Limanda yokohamae*. Bulletin of the Japanese Society of Scientific Fisheries, 46(4):499-503.

Fukusho, K. 1983. Present status and problems of the rotifer *Brachionus plicatilis* for fry production of marine fishes in Japan. Advances and Perspectives in Aquaculture. Coquimbo, Chile, Universidad del Norte. 361-374.

Fukusho, K. and K. Hirayama (eds). 1992. The First Live Feed - *Brachionus plicatilis*. Honolulu, HI, The Oceanic Institute.

Fukusho, K. and M. Okauchi 1982. Strain and size of the rotifer, *Brachionus plicatilis*, being cultured in Southeast Asian countries. Bull. Natl. Res. Inst. Aquaculture 3: 107-109.

Fulks, W. and K. L. Main (eds.) 1991. Rotifer and Microalgae Culture Systems. Honolulu, HI, The Oceanic Institute.

Furukawa, I. and K. Hidaka 1973. Technical problems encountered in the mass culture of the rotifer using marine yeast as food organisms. Bulletin Plankton Society Japan 20:61-71.

Furuta, S., T. Watanabe, H. Yamada, T. Miyanaga. 1997. Changes in feeding condition of released hatchery-reared Japanese flounder, *Paralicthys olivaceus*, and prey mysid density in the coastal area of Tottori prefecture. Nippon Suisan Gakkaishi 63:886-891.

Galtsoff, P.S., F. E. Lutz, P. S. Welch, and J. G. Needham. 1937. Culture Methods for Invertebrate Animals. Comstock Pub. Co., Inc., Ithaca, N.Y. 590pp.

Gatesoupe, F. J. 1982. Nutritional and anti-bacterial treatments of live food organisms: the influence on survival, growth rate and weaning success of turbot (*Scophthalmus maximus*). Annals Zootechnology 31:353-368.

Gatesoupe, F. J. 1991. The effect of three strains of lactic bacteria on the production rate of rotifers, *Brachionus plicatilis*, and their dietary value for turbot larvae, *Scophthalmus maximus*. Aquaculture 96: 335-342.

Gatesoupe, F. J. 1994. Lactic acid bacteria increase the resistance of turbot larvae, *Scophthalmus maximus*, against pathogenic vibrio. Aquat. Living Resources 7: 277-282.

Gatesoupe, F. J. and Le Millnaire C., 1984. Dietary value adaptation of live food organisms for covering the nutritional requirements of marine fish larvae: 231. In: Acts of Norwiegian, French shop on Aquaculture in Brest, France, INFREMER (Ed.), 253pp.

Gealy, D. 2003. Keeping and breeding Australian rainbow fish. Web contact. http://rainbowfishes.org/index.html

Geiger, J.G. 1983. A review of pond zooplankton production and fertilization for the culture of larval and fingerling striped bass. Aquaculture 35:353-369.

Gladue, R. 1991. Heterotrophic microalgae production potential for application to aquaculture feeds. In: Rotifer and Microalgae Culture Systems. The Oceanic Inst., Hawaii. pp 275-286.

Goldman, J. C. 1979. Outdoor algal mass culture - I applications. Water Research, Vol. 13, pp 1-19.

Gómez A, M. Serra, G.R. Carvalho, & D.H. Lunt, 2002. Speciation in ancient cryptic species complexes: evidence from the molecular phylogeny of *Brachionus plicatilis* (Rotifera). Evolution 56:1431-1444

Gonzalez-Rodriguez, E. and S. Y. Maestrini 1984. The use of some agricultural fertilizers for the mass production of marine algae. Aquaculture, Vol. 36, pp 245-256.

Good, L.K., R.C. Bayer, M.L. Gallagher, J.H. Rittenburg. 1982. Amphipods as a potential diet for juveniles of the American lobster *Homarus americanus* (Milne Edwards). J. Shellfish Res. 2:183-187.

Gordo, T., L. M. Lubian and J. P. Canavate 1994. Influence of temperature on growth, reproduction and longevity of *Moina salina* Daday, 1888 (Cladocera, Moinidae). J. Plankton Res. 16:1513-1523.

Grabner, M., W. Wieser, and R. Lackner. 1981. The suitability of frozen and freeze-dried zooplankton as food for fish larvae: A biochemical test program. Aquaculture 26(1-2): 85-94.

Groenweg, J. and M. Schluter 1981. Mass production of freshwater rotifers on liquid wastes. II. Mass production of *Brachionus rubens* Ehernberg 1838 in the effluent of high-rate algal ponds used for the treatment of piggery waste. Aquaculture 25:25-33.

Hadley, N. H., J. J. Manzi, A. G. Eversole, R. T. Dillon, C.E. Battey and N. M. Peacock. 1997. A manual for the culture of the Hard Clam, *Mercenaria spp.* in South Carolina. Pub. Sea Grant Consortium, Charleston S.C. 135 pgs.

Hagiwara, A., A. Hino, 1989. Effect of incubation and preservation on resting eggs hatching and mixis in the derived clones of the rotifer Brachionus plicatilis. Hydrobiology 186(9):415-421.

Hagiwara, A. and K. Hirayama 1993. Preservation of rotifers and its application in the finfish hatchery. TML Conference Proceedings 3: 61-71.

Hagiwara, A. 1994. Practical use of rotifer cysts. Israel J. Aquaculture 46: 13-21.

Hagiwara, A., C. S. Lee and D. J. Shiraishi 1995b. Some reproductive characteristics of the broods of the Harpacticoid copepod *Tigriopus japonicus* cultured in different salinities. Fisheries Science 61:618-622.

Hagiwara, A., M.-M. Jung, T. Sato and K. Hirayama 1995. Interspecific relations between marine rotifer *Brachionus rotundiformis* and zooplankton species contaminating in the rotifer mass culture tank. Fisheries Science 61: 623-627.

Hagiwara, A., N. Hoshi, F. Kawahara, K. Tominaga and K. Hirayama 1995. Resting eggs of the marine rotifer *Brachnious plicatilis* Muller: development, and effect of irradiation on hatching. Hydrobiologia 313/314: 223-229.

Hagiwara, A., W.G. Gallardo, M. Assavaaree, T. Kotani, and A.B. de Araujo. 2001. Live food production in Japan: Recent progress and future aspects. Aquaculture 200:111-127.

Hagiwara, A., K. Suga, A. Akazawa, T. Kotani and Y. Sakakura. 2007. Development of rotifer strains with useful traits for rearing fish larvae. Aquaculture 268:44-52.

Hamada, K. and A. Hagiwara 1993. Use of preserved diet for rotifer *Brachionus plicatilis* resting egg formation. Nippon Suisan Gakkaishi 59: 85-91.

Hamre, K., A. Srivastava, I Ronnestad, A Mangor-Jensen, and J. Stoss. 2008. Several micronutrients in the rotifer Brachionus sp. may not fulfil the nutrient requirements of marine fish larvae. Aquaculture Nutrition 14:51-60.

Hanazato, T. and M. Yasuno. 1984. Growth, reproduction, and assimilation of *Moina macrocopa* fed on *Microcystis* and/or *Chlorella*. Japanese Journal Ecology 34(2): 195-202.

Hanlon, R.T., Turk, P.E., & Lee, P.G. 1991. Squid and cuddlefish mariculture: and updated perspective. Journal of cephalopod Biology, 2, 31-40.

Heinle, D.R. and D.A. Flemmer, 1975. Carbon requirements of a population of the estuarine copepod *Eurytemora affinis*. Mar. Biol. 31:235-247.

Heisig, G. 1977. Mass cultivation of *Daphnia pulex* in ponds: The effects of fertilization, aeration, and harvest on the population development. In: Cultivation of Fish Fry and Its Live Food, E. Styczynska-Jurewicz, T. Backiel, E. Jaspers, and G. Persoone (eds.). European Mariculture Society Special Publication 4: 335-359.

Helfrich, P. 1973. The feasibility of brine shrimp production on Christmas island. Sea Grant Technical Report UNIHI - Sea grant TR-73-02, 1-173.

Hemdal, J. 2000. Raising Mysid Shrimp as a Home Aquarium Food. Seascope, Aquarium Systems.

Henzelmann, E. 1960. Micro Worm Culture - New Method. Aquatic Life and Aquatic World, March-April Issue Vol.X#2.

Herald, E.S. and R.P. dempster. 1955. Brine shrimp vs. copper solutions. Aquarium Journal, 35:334.

Herrero, C., C. Angeles, J. Fabregas and J. Abalde 1991. Yields in biomass and chemical constituents of four commercially important marine microalgae with different culture media. Aquaculture Engineering 10: 99-110.

Hills, C. and H. Nakamura 1978. Food from sunlight. Planetary survival for hungry people. University of Trees Press, Boulder Creek, CA. ISBN 0-916438-13-9.

Hino, A. 1993. Present culture systems of the rotifer (*Brachionus plicatilis*) and the function of microorganisms. TML Conference Proceedings 3: 51-59.

Hirata, H. 1979. Rotifer culture in Japan. In: Cultivation of Fish Fry and Its Live Food, E. Styczynska-Jurewicz, T. Backiel, E. Jaspers and G. Personne (eds.) European Mariculture Society, Special Publication 4:361-375.

Hirata, H. S. Yamasaki, T. Kawaguchi and M. Ogawa. 1983. Continuous culture of the rotifer *Brachionus plicatilis* fed recycled algal diets. Hydrobiologia 104:71-75.

Hirata, H., O. Murata, S. Yamada, H. Ishitani, and M. Wachi. 1998. Probiotic culture of the rotifer *Brachionus plicatilis*. Hydrobiologia 387:495-498.

Hirayama, K. and A. Hagiwara 1995. Recent advances in biological aspects of mass culture of rotifers (*Brachionus plicatilis*) in Japan. ICES Mar. Sci. Symp. 201: 153-158.

Hirayama, K. and C. G. Satuito 1991. The nutritional improvement of baker's yeast for the growth of the rotifer *Brachionus plicatilis*. In: Rotifer and Microalgae Culture Systems. The Oceanic Institute, Hawaii . pp 151-162.

Hirayama, K. and H. Funamoto. 1983. Supplementary effect of several nutrients on nutritive deficiency of baker's yeast for population growth of the rotifer *Brachionus plicatilis*. Bulletin Japanese Society Scientific Fisheries 49:505-510.

Hirayama, K. and S. Ogawa 1972. Fundamental studies on the physiology of the rotifer for its mass culture. I. Filter feeding of rotifer. Bulletin Japanese Society Scientific Fisheries 38:1207-1214.

Hirayama, K. and T. Kusano 1972. Fundamental studies on the physiology of rotifer for its mass culture. II. Influence of water temperature on the population growth of rotifer. Bulletin Japanese Society Scientific Fisheries 38:1357-1363.

Hirayama, K. 1987. A consideration of why mass culture of the rotifer *Brachionus plicatilis* with baker's yeast is unstable. Hydrobiologia 147:269-270.

Hirayama, K. 1990. A physiological approach to problems of mass culture of the rotifer. 73-80.

Hirayama, K., I. Maruyama and T. Maeda 1989. Nutritional effect of freshwater *Chlorella* on growth of the rotifer *Brachionus plicatilis*. Hydrobiologia 186/187: 39-42.

Hirayama, K., K. Watanabe and T. Kusano. 1973. Fundamental studies on the physiology of rotifer for its mass culture. III. Influence of phytoplankton density on population growth. Bulletin Japanese Society Scientific Fisheries 39:1123-1127.

Hiyama, Y. and R.P. Singh. 1966. Quantitative relationship between marine and freshwater preys and predators as determined by S35. Bull. Jap. Soc. Sci. Fish., 32:705.

Hoff, F. and D. Wilcox. 1986. Unpublished data from a *Moina* culture project at VW Tropical Fish, Florida from 1986 to 1991. Aquaculture Consultants, Dade City, FL.

Hoff, F. and T.W. Snell 1988. A New Live Food for Tropical Fish, Special Publication, Florida Department of Agriculture.

Hoff, F. H. Jr, 1996. Conditioning, spawning, and culture of marine tropical fish, with emphasis on *Amphiprion* (clownfish) species. Florida Aqua Farms, Dade City, Florida.

Højgaard J.K., Per M Jepsen & Benni W Hansen, 2008. Salinity-induced quiescence in eggs of the calanoid copepod *Acartia tonsa* (Dana): a simple method for egg storage. Department of Environment, Social and Spatial Change, Roskilde University, Universitetsvej, Roskilde, Denmark. Published by Blackwell Publishing Ltd.

Houde, E. D. and C. E. Zastrow 1993. Ecosystem- and taxon-specific dynamic and energetics properties of larval fish assemblages. Bull. Marine Sci. 53: 290-335.

Houde, E.D. and R.C. Scheker 1980. Feeding by marine fish larvae: developmental and functional responses. Environmental Biology of Fish 5:315-334.

Howell, B.R. 1972. Preliminary experiments on the rearing of lemon sole *Microstomus kitt* (Walbaum) on cultured foods. Aquaculture 1:39-44.

Huner, Jay V. and E. E. Brown (Editors). 1985. Crustacean and mollusk aquaculture in the United States. Avi Publishing Company, Westport CT. Book 476 pgs.

Hunter, J..R. 1981. The essentials of brine shrimp. Farm Pond Harvest. Fall issue, pp17-18.

Hunter, J.R. 1972. Swimming and feeding behavior of larval anchovy, *Engraulis mordax*. Fishery Bulletin U.S. 70:821-838.

Hunter, J.R. 1976. Culture and growth of the northern anchovy, *Engraulis mordax*, larvae. Fisheries Bulletin U.S. 74:81-88.

Hunter, J.R. 1980. The feeding behavior and ecology of marine fish larvae. In: Fish Behavior and its Use in the Capture and Culture of Fishes, J.E. Bardach, J.J. Magnuson, R.C. May and J.M. Reinhart (eds), ICLARM Conference Proceedings 5:287-330.

Intriago, P. and D. A. Jones 1993. Bacteria as food for *Artemia*. Aquaculture 113: 115-127.

Ito,T. 1960. On the culture of the mixohaline rotifer *Brachionus plicatilis* O.F. Muller in the seawater. Report Faculty Fisheries, Prefectural University of Mie 3:708-740.

Ivleva, I.V. 1969. Mass Cultivation of Invertebrates, Biology and Methods. Academy of Sciences of the U.S.S.R., All-Union Hydrobiological Society: 148 pp. (Transl. from Russian by A. Mercado 1973).

Iwai, T. 1980. Sensory anatomy and feeding of fish larvae. In: Fish Behavior and its Use in the Capture and Culture of Fishes, J.E. Bardach, J.J. Magnuson, R.C. May and J.M. Reinhart (eds), ICLARM Conference Proceedings 5:124-145.

James, C. M. and T. Abu-Rezeq 1989a. An intensive chemostat culture system for the production of rotifers in aquaculture. Aquaculture 81: 291-301.

James, C. M. and T. Abu-Rezeq 1989b. Intensive rotifer cultures using chemostats. Hydrobiologia 186/187: 422-430.

James, C. M., A. M. Al-Khars, et al. 1988. pH dependent growth of *Chlorella* in a continuous culture system. J. World Aquacult. Soc. 19: 27-35.

James, C.M., P. Dias and A.E. Salman 1987. The use of marine yeast (*Candida* sp.) and baker's yeast (*Saccharomyces cerevisiae*) in combination with *Chlorella* sp. for mass culture of the rotifer *Brachionus plicatilis*. Hydrobiologia 147:263-268.

Johnston, R. 1963. Seawater, the natural medium of phytoplankton. I. General features. Journal Marine Biological Association of U.K. 43:409-425.

Jones, A. 1986. Live verses inert feeds in larval aquaculture: Panel discussion. World Aquaculture Society Meeting, Reno, Nevada.

Kabatashi, M. 1981. Respiratory function of hemoglobin in *Moina macrocopa*. Comp. Biochem. Physiol. A Comp Physiol. 70(3): 381-386.

Kahan, D., G. Uhlig, D. Schwenzer and L. Horowitz 1981. A simple method for cultivating harpacticoid copepods and offering them to fish larvae. Aquaculture 26: 303-310.

Kain, J.M. and G.E. Fogg. 1958. Studies on the growth of marine phytoplankton. I. *Asterionella japonica* Gran. Journal Marine Biological Association of U.K. 37:397-413.

Kessler, E. 1976. Comparative physiology, Biochemistry and the taxonomy of *Chlorella* (Chlorophyceae). Plant Systematics and Eolution 125: 125-138.

Khanaichenko, A.N. 1998. Approach to optimize copepod cultures exploitation. Poster at the Aquaculture Europe Meeting 1998 Bordeaux, France. (found on the web).

King, C.E. and T.W. Snell 1977. Sexual recombination in rotifers. Heredity 39:357-360.

Kinne, O. 1977. Cultivation of animals: axenic cultivation. In: O. Kinne (Editor), Marine Ecology, Vol. 3, Part 3, Wiley Interscience, New York, NY, pp. 1295-1314.

Kitajima, C., S. Fujita, F. Oowa, Y. Yone and T. Watanabe 1979. Improvement of dietary value for red sea bream larvae of rotifers, *Brachionus plicatilis*, cultured with baker's yeast, *Saccharomyces cerevisiae* . Bulletin Japanese Society Scientific Fisheries 45:469-471.

Kitajima, C., K. Fukusho, H. Iwamoto, H. Yamamoto, 1976. Amount of rotifers, *Brachionus plicatilis*, consumed by red sea bream larvae, *Pagrus major*. Bull. Nagaski Pref. Inst. Fish., 1:105-112.

Klee, A. J. 1959. Beginners Corners on Micro Worms. Aquarium Journal, San Francisco Aquariun Society, Vol. #4.

Knaack, J. 1958. Mikro-Alchem.-Die Gattung Turbatrix Aquarien und Terrarien, No. 1.

Komis, A.. 1992. Improved production and utiliization of the rotifer *Brachionus plicatilis* Muller, in european sea bream (*Sparus aurata* Linnaeus) and sea bass (*Dicentrarchus labrax* Linnaeus) larviculture. Thesis. University of Gent, Belgium.

Korstad, J., A. Neyts, T. Danielsen, I. Overrein and Y. Olsen 1995. Use of swimming speed and egg ratio as predictors of the status of rotifer cultures in aquaculture. Hydrobiologia 313/314: 395-398.

Koste, W. 1980. Das Radertier-Portrat: *Brachionus plicatilis*, ein salzwasserradertier. Mikrokosmos 5:148-155.

Kostopoulou, V. and O. Vadstein. 2007. Growth performance of the rotifers *Brachionus plicatilis*, B. 'Nevada' and B. 'Cayman' under different food concentrations. Aquaculture 273:449-458.

Kraul, S. 1981. Personal communication. Unpublished data. Waikiki Aquarium, Hawaii..

Kuhlmann, D., G. Quantz, and U. Witt. 1981. Rearing of turbot larvae (Scophthalmus maximus L.) on cultured food organisms and postmetamorphosis growth on natural and artificial food. Aquaculture 23:183-196.

Lamm. D. R. 1987. Culturing copepods, a food for marine fish larvae. Freshwater and Marine Aquarium Mag.

Langis, R., D. Proulx, J. de la Nouc, and P. Couture. 1988. Effects of a bacterial biofilm on intensive *Daphnia* culture. Aquacultural Engineering 7: 21-38.

Lavens, P. and P. Sorgeloos. 1996. Manual on the production and use of live food for aquaculture. FAO Fisheries Technical Paper. No. 361. Rome, FAO. 295p.

Lee, C.S. and F. Hu 1980. Salinity tolerance and salinity effect on brood size of *Tigriopus japonicus* Mori. Aquaculture 22: 377-381.

Lee, J. J., Hunter, S. H. and Bovee, E. C. (editors). 1985. An illustrated Guide to the Protozoa, Society of Protozoologists, Lawrence, Kansas.

Lee Heng, L.J. 1983. Preliminary studies of the culture of *Moina* using organic wastes. Dept. Zoology, Nat. Univ. of Sinapore.

Leger, P. and P. Sorgeloos. 1982. Automation in stock culture maintenance and juvenile separation of the mysid *Mysidopsis bahia* (Molenock). Aquacult. Engineering 1:45-53.

Leger, P., P. Vanhaecke and P. Sorgeloos 1983. International Study on *Artemia*. XXIV. Cold storage of live *Artemia* nauplii from various geographical sources: Potentials and limits in aquaculture. Aquacult. Engineering 2: 69-78.

Leighton, T.. 1981. Production of live foods for tropical marine larval fishes. Personal communication, unpublished data. Waikiki Aquarium, Honolulu, Hawaii.

Lemus, J.T., J.T. Ogle and J.M. Lotz, 2004. Increasing Production of Copepod Nauplii in a Brown-Water Culture with Supplemental Feeding. North American Journal of Aquaculture, (Vol. 66) (No. 3) 169-176

Li, Y., S. Jin, and J. Qin. 1996. Strategies for development of rotifers as larval fish food in ponds. J. World Aquaculture Soc. 27:178-186.

Lincoln, E.P., T.W. Hall, and B. Koopman. 1983. Zooplankton control in mass algae cultures. Aquaculture 32: 331-337.

Loedolff, C.L. 1964. Function of Cladocera in oxidation ponds. Journal Water Pollution Control Federation: 36(3).

Londsdale, D., D. Heinle, C. Siegfried 1979. Carnivorous feeding behavior of the adult calanoid copepod *Acarti tonsa*, Dana. Exper. Mar. Biol. and Ecol. 36:235-248.

Lubzens, E. 1981. Rotifer resting eggs and their application to marine aquaculture. European Mariculture Society, Special Publication 6:163-179.

Lubzens, E. 1987. Raising rotifers for use in aquaculture. Hydrobiologia 147:245-255.

Lubzens, E. 1988. Possible use of rotifer resting eggs and preserved live rotifers (*Brachionus plicatilis*) in aquaculture and mariculture. European Aquaculture Society, Special Publication.

Lubzens, E., A. Tandler, and G. Minkoff. 1989. Rotifers as food in aquaculture. Hydrobiologia 186/187:387-400.

Lubzens, E., A. Marko and A. Tietz 1984. Lipid synthesis in the rotifer *Brachionus plicatilis* . European Mariculture Society, Special Publication 8:201-210..

Lubzens, E., G. Kolodny, et al. 1990. Factors affecting survival of rotifers (*Brachionus plicatilis* O.F. Muller) at 4°C. Aquaculture 91: 23-47.

Lubzens, E., G. Minkoff and S. Marom 1985. Salinity dependence of sexual and asexual reproduction in the rotifer *Brachionus plicatilis* . Marine Biology 85:123-126.

Lubzens, E., O. Gibson, O. Zmora and A. Sukenik 1995. Potential advantages of frozen algae (*Nannochloropsis* sp.) for rotifer (*Brachionus plicatilis*) culture. Aquaculture 133: 292-309.

Lubzens, E., O. Zmora, and Y. Barr. 2001. Biotechnology and aquaculture of rotifers. Hydrobiologia 446:337-353.

Ludwig, G. M. 1994. Rearing sunshine bass fry on tank-cultured freshwater rotifers. Aquaculture Magazine 20: 68-70.

Ludwig, G.M. 1997. Sunshine bass fry culture. Aquaculture Magazine 23:93-97.

Lund, J.W.G. 1965. The ecology of the freshwater phytoplankton, Biological Review, 40, 231-293.

Lussier, S.M., A. Kuhn, M.J. Chammas, and J. Sewall. 1988. Techniques for the laboratory cutlure of *Mysidopsis* species (Crustacea: Mysidacea). Env. Toxicol. Chem. 7:969-977.

Lutz, F. E., *et al*, 1937. Culture Methods for Invertebrate Animals, Dover Publications, Inc. New York, N.Y.: 590 pp.

Maeda, B. and A. Hino, 1991. Environmental management for mass culture of the rotifer, *Brachionus plicatilis*.. Rotifer and Microalgae Culture Systems. The Oceanic Inst., Hawaii. pp 125-133.

Makridis, P., A.J. Fjellheim, J. Skjermo, and O. Vadstein. 2000. Control of bacterial flora of *Brachionus plicatilis* and *Artemia franciscana* by incubation in bacterial suspensions. Aquaculture 185:207-218.

Marcus, N. H. and F. Boero 1998. The Nature of Resting Stages and their Role in the Population Dynamics of Marine and Freshwater. Published in Limnology and Oceanography, 1998 Crustaceans.

Marcus, N.H. and M. Murray. 2001. Copepod diapause eggs: a potential source of nauplii for aquaculture. Aquaculture 201:107-115.

Marini, F. and D. Sapp (2003). Copepod Culture. The breeders net http://www.advancedaquarist.com/issues/fed2003/breeder2.htm

Maruyama, I. and K. Hirayama 1993. The culture of the rotifer *Brachionus plicatilis* with *Chlorella vulgaris* containing vitamin B$_{12}$ in its cells. J. World Aquaculture Soc. 24: 194-198.

Maruyama, T. Nakamura, et al. 1986. Identification of the alga "marine Chlorella" as a member of Eustigmatophyceae. Jpn. J. Phycol. 4: 319-325.

Mathais, J.A. and M. Papst. 1981. Growth, survival and distribution of *Gammarus lacustris* (Crustacea-Amphipoda) stocked into ponds. Can. Tech. Rep. Fish. Aquat. Sci. 989: 1-15.

Mauchline, J. 1980. The biology of mysids and euphausids. In: Advances in Marine Biology. Part 1. The Biology of Mysids, Vol. 18, Academic Press, London.

May, R.C., R.S.V. Pullin, and V.G. Jhingran (Eds.). 1984. Summary Report of the Asian Regional Workshop on Carp Hatchery and Nursery Technology. Manila, Philippines, 1-3 February, 1984: 38pp.

McKinnon, A.D., S. Duggan, P.D. Nichols, M.A. Rimmer, G. Simmons, and B. Robino. 2003. The potential of tropical paracalinid copepods as live feeds in aquaculture. Aquaculture 223: 89-106.

Menzel, R. W. 1971. Quahog clams and their possible mariculture. Proc. World Mariculture Soc. Second Annual Meeting. Pgs 23-36

Merchie, G., P. Lavens, Dhert Ph., M. Deshasque, H. Nelis, A. De-Leenheer, & P. Sorgeloos. 1995. Variation of ascorbic acid content in different live food organisms. Aquaculture, 134(3-4):325-337.

Minkoff, G., E. Lubzens and E. Meragelman 1985. Improving asexual reproduction rates in a rotifer (*Brachionus plicatilis*) by salinity manipulations. Israel Journal Zoology 33:195-203.

Misra, S. K. and R. P. Phelps 1992. A zooplankton harvester designed to colllect rotifers. Prog. Fish. Culturist 54: 267-269.

Mitchell, S. A. 1992. The effect of pH on *Brachionus calyciflorus* Pallas (Rotifera). Hydrobiologia 245: 87-93.

Miyakawa, M. and K. Muroga 1988. Bacterial flora of cultured rotifer *Brachionus plicatilis*. Suisan-zoshoku 35: 237-243.

Moe, M. A. 1989. The Marine Aquarium Reference, Systems and Invertebrates. Green Turtle Publications, Plantation FL,

Monakov, A.V. 1972. Review of studies on feeding of aquatic invertebrates conducted at the Institute of Biology of Inland Waters, Academy of Science, U.S.S.R. Journal Fisheries Research Board of Canada 29(4): 363-383.

Morton, R. and E. Cook, 1966. Nematode Biochemistry VI. Conditions for Axenic Culture of *Turbatrix aceti*; *Pangrellus redivivus*, *Rhabditis anomala*, and *Caenorhabditis briggsae*. Comp. Biochem. Physiol., Vol. 17, No. 2.

Mullin, M. M. and E. R. Brooks. 1967. Laboratory culture, growth rate, and feeding behavior of a planktonic marine copepod. Limnol. Oceanogr 12, 657-666.

Muroga, K., M. Higashi and H. Keitoku 1987. The isolation of intestinal microflora of farmed red sea bream (*Pagrus major*) and black sea bream (*Acanthopagrus schlegeli*) at larval and juvenile stages. Aquaculture 65: 79-88.

Museum 1996. Amphipod Culture. Web site of the Museum Victoria Australia http//www.mov.vic.gov.au/crust/page1a. html

Nagata, W. D. and J. N. C. Whyte 1992. Effects of yeast and algal diets on the growth and biochemical composition of the rotifer *Brachionus plicatilis* (Muller) in culture. Aquacult. Fish. Management 23: 13-21.

Nellen, W., Quantz, G. Witt, U., Kuhlmann, D. and Koske, P.H. 1981. Marine fish rearing on the base of an artificial food chain. In: European Mariculture Society, Special Publication No. 6, Bredene, Belgium, pp 133-147.

O'Brien, W.J. 1979. The predator-prey interaction of planktivorous fish and zooplankton. American Scientist 67:572-581.

O'Brien, W.J. 1987. Planktivory by freshwater fish: thrust and parry in pelagia. In: Predation: Direct and Indirect Impacts on Aquatic Communities, pgs. 3-16, C.W. Kerfoot and A. Sih (eds.), University Press of New England, Hanover, New Hampshire.

O'Kelly, J.C. 1974. Inorganic nutrients. In: Algal Physiology and Biochemistry, W.D.P. Stewart (ed.), Botanical Monographs Vol. 10, Univ. Calif. Press, Berkeley, pgs. 610-635.

Ogle, J. 1979. Adaption of a brown water culture technique to the mass culture of the copepod *Acartia tonsa*. Gulf Res. Report 6 No. 3,000-000.

Ogle, J., C. Nicholson and J. M. Lotz. (2002). Culture of the copepod *Acartia tonsa* utilizing various artificial foods. Gulf Caribbean Fisheries Meeting 53 Annual Meeting, Year 2000. pgs. 234-240.

Ohno, A. and Y. Okamura 1988. Propagation of the calanoid copepod, *Acartia tsuensis*, in outdoor tanks. Aquaculture 70: 39-51.

Oka, A., N. Suzuki, and T. Watanabe. 1982. Effects of ω3 fatty-acids in *Moina* on the fatty-acids composition of larval aye, *Plecoglossus altivelis*. Bulletin Japanese Society Scientific Fisheries 48(8): 1159-1162. (In Japanese, English summary)

Okauchi, M. and K. Fukusho 1984a. Environmental conditions and medium required for mass culture of Prasinophyceae, *Tetraselmis tetrathele*. Bulletin National Research Institute Aquaculture 5:1-11.

Okauchi, M. and K. Fukusho 1984b. Food value of the minute algae, *Tetraselmis tetrathele* for the rotifer *Brachionus plicatilis*. I. Population growth with batch culture Bulletin National Research Institute Aquaculture 5:13-18.

Okauchi, M. 1991. The status of phytoplankton production in Japan. In: Rotifer and Microalgae Culture Systems. The Oceanic Institute, Hawaii. pp 247-256.

Oltra, R. and R. Todoli. 1997. Effects of temperature, salinity and food-level on the life-history traits of the marine rotifer *Synchaeta cecilia valentina* N. subsp. J. Plankton Res. 19:693-702.

Oltra, R., R. Todoli, T. Bosque, L.M. Lubian, and J.C. Navarro. 2000. Life history and fatty acid composition of the marine rotifer *Synchaeta cecilia valentina* fed different algae. Mar. Ecol. Prog. Ser. 193:125-133.

Paffenhofer, G. and R. P. Harris. 1979. Laboratory culture of marine planktonic food webs. Advances in Marine Biol., 61:211-308.

Palmer, C. M. 1977. Algae and Water Pollution. EPA-600/9-77-036. U.S. Environmental Protection Agency. Cincinnati, OH. 124 pp.

Papakostas, S., S. Dooms, A. Triantafyllidis, D. Deloof, I. Kappasa, K. Dierckens, T. De Wolf, P. Bossier, O. Vadstein, S. Kui, P. Sorgeloos and T.J. Abatzopoulos. 2006. Evaluation of DNA methodologies in identifying Brachionus species used in European hatcheries. Aquaculture 255:557-564.

Park, H.G., K.W. Lee, H.S. Kim, and M.M. Jung. 2001. High density culture of the freshwater rotifer, *Brachionus calyciflorus*. Hydrobiologia 446/447:369-374.

Pascual, E. and M. Yufera 1983. Crecimento en cultivo de una cepa de *Brachionus plicatilis* O.F. Muller en function de la temperatura y la salinidad. Investigationes Pesquera 47:151-159.

Patrick, R. and Reimer, C. W. 1966. The Diatoms of the United States exclusive of Alaska and Hawaii. Vol. 1. Fragilariaceae, Eunotiaceae, Achnanthaceae, Naviculaceae. Monogr. of The Acad. of Natural Sci. of Philadelphia, No. 13, 213 pp.

Pennak, R. W., 1953. Freshwater Invertebrates of the United States. Ronald Press, New York, pp 321-469.

Pavlyutin, A.P. 1976. Food value of detritus for some freshwater cladocera species. Gidrobiol Zh 12(4): 15-21. (In Russian, English summary)

Persoone, G., P. Sorgeloos, O. Roels and E. Jaspers 1980. The Brine Shrimp *Artemia*. Volume 3, Ecology, Culturing and Use in Aquaculture. Universa Press, Wettern, Belgium.

Pilarska, J. 1972. The dynamics of growth of experimental populations of the rotifer *Brachionus rubens* Ehrbg. Polish Archives Hydrobiology 19:265-277.

Pilarska, J. 1977. Ecophysiological studies on *Brachionus rubens* Ehrbg (Rotatoria) I. Food selectivity and feeding rate. Polish Archives Hydrobiology 24:319-328.

Piparo, A. J. 1979. Culturing Microworms. Freshwater and Marine Magazine. Vol. 2 #1 January.

Planas, M., J.A. Vázquez, J. Marqués, R. Pérez-Lomba, M.P. González and M. Murado. 2004. Enhancement of rotifer (*Brachionus plicatilis*) growth by using terrestrial lactic acid bacteria. Aquaculture 240:313-329.

Polo, A., M. Yufera, et al. 1992. Feeding and growth of gilthead seabream (Sparus aurata L.) larvae in relation to the size of the rotifer strain used as food. Aquaculture 103: 45-54.

Poulet, S.A. 1977. Grazing of marine copepods development stages on naturally occurring particles. Jour. Fish. Res. Bd. Can 34(12):2381-2387.

Poulet, S.A. 1978. Comparison between five coexisting species of marine copepods feeding on naturally occuring particulate matter. Limnology and Oceanography 23:1126-1143

Poulet, 1983. Factors controlling utilization of non-algal diets by particle grazing copepods: a review. Oceanologica Acta 69(3): 221-234.

Pourriot, R. and T.W. Snell 1983. Resting eggs in rotifers. Hydrobiologia 104:213-224.

Pourriot, R., C. Rougier and D. Benest 1987. Qualite de la nourriture et controle de la mixis chez le rotifere *Brachionus rubens*, Ehrbg. Bulletin de la Societe Zooglque de France 111:105-111.

Pratt, F. and J. Fong. 1940. Studies on *Chlorella vulgaris*. II. Further evidence that *Chlorella* cells form a growth-inhibiting substance. Amer. J. Bot. 27:431-436.

Prescott, G. W. 1962. Algae of the Western Great Lakes Area. Wm. C. Brown Co., Dubuque, Iowa. 977 pp.

Prescott, G. W. 1970. How to Know the Freshwater Algae. Wm. C. Brown Co., Dubuque, Iowa. 348 pp.

Price, K.S., Jr., W.N. Shaw, K.S. Danberg, 1976. First International Conference of

Price, W.W. 1982. Key to the shallow water Mysidacea of the Texas coast with notes on their ecology. Hydrobiologia 93:9-21.

Price, W.W., R.W. Heard, and L. Stuck. 1994. Observations on the genus *Mysidopsis* Sars, 1864 with designation of a new genus *Americamysis*, and the descriptions of *Americamysis alleni* and *A. stucki* (Pericarida: Mysidacea: Mysidae), from the Gulf of Mexico. Proc. Biol. Soc. Washington 107:680-698.

Provasoli, L. 1968. Media and prospects for the cultivation of marine algae. In : A Watanabe and A. Hatteri (Eds), Cultures and collection of algae. Pro. U. S. Japan Conf. Hakone, Sept 1966. Jap. Soc. Pl. Phy. Siol.. pp 63-75.

Pryor, V. K. and C. E. Epifanio 1993. Prey selection by larval weakfish (*Cynoscion regalis*): the effects of prey size, speed, and abundance. Marine Biol. 116: 31-37.

Rao, M.V. and K.P. Krishnamoorthi. 1977. Preferential devouring of blue-green algae by a daphnia *Moina dubia*. Indian Journal Environmental Health 19(2): 143-144.

Raymont, J.E.G. and R.S. Miller 1962. Production of marine zooplankton with fertilization in an enclosed body of sea water. Hydrobiol. 47:169-209.

Reddy, S.R. and K. Shakuntala. 1986. Use of mysids as food for culture of juvenile *Penaeus merguiensis*. In: Biology of Benthic Marine Organisms: Techniques and Methods Applied to the Indian Ocean, pgs. 359-364, Indian Ed. Serv., no. 12.

Schluter, M. and J. Groeneweg. 1981. Mass production of freshwater rotifers on liquid wastes. I. The influence of some environmental factors on population growth of *Brachionus rubens* Ehrenberg 1838. Aquaculture 25:17-24.

Reeve, M.R. 1963. Growth efficiency in *Artemia* under laboratory conditions. Biological Bulletin 125:133-145.

Reguera, B. 1984. The effect of ciliate contamination in mass cultures of the rotifer *Brachionus plicatilis* O.F. Muller. Aquaculture 40:103-108.

Reitsema, L.A. and J.M. Neff. 1980. A recirculating artificial seawater system for the laboratory culture of *Mysidopsis almyra* (Crustacea; Pericaridea). Estuaries 3:321-323.

Ribblett S.G. and D.W. Coats, 2000. Life Among Drowning Leaves. Smithsonian Environmental Research Center Newsletter. Spring 2000.

Rico-Martinez, R. and S. I. Dodson 1992. Culture of the rotifer *Brachionus calyciflorus* Pallas. Aquaculture 105: 191-199.

Rothbard, S. 1975. Control of *Euplotes* sp. by formalin in growth tanks of *Chlorella* sp. which serves as food for hatchlings. Bamidgeh 4:100-109.

Rottmann, R.W. and J.V. Shireman. 1989. Hatchery manual for grass carp and other Chinese carp. IFAS Extension Circular. University of Florida. Gainesville, FL

Rottmann, R.W. 1992. Culture Techniques of *Moina*: The Ideal Daphnia for Feeding Freshwater Fish Fry IFAS Extension Circular 1054, University of Florida. Gainesville, FL

Roman, M.R. 1977. Feeding of the copepod *Acartia tonsa* on the diatom *Nitzschia closterium* and brown algae (*Fucus vesiculosus*) detritus. Mar. Biol. 42:149-155.

Round, F.E. 1981. The Ecology of Algae. Cambridge University Press, New York.

Schluter, M. 1980. Mass culture experiments with *Brachionus rubens* . Hydrobiologia 73:45-50.

Scott, A.P. and S.M. Baynes 1978. Effect of algal diet and temperature on the biochemical composition of the rotifer *Brachionus plicatilis* . Aquaculture 14:247-260.

Scott, J. M. 1981. The vitamin B12 requirement of the marine rotifer *Brachionus plicatilis*. J. Mar. Biol. Assoc. U. K. 61: 983-994.

Segawa, S. and W. T. Yang 1987. Reproduction of an estuarine *Diaphanosoma aspinosum* (Branchiopoda: Cladocera) under different salinities. Bull. Plankton Soc. Japan 34: 43-51.

Segawa, S. and W. T. Yang 1988. Population growth and density of an estuarine cladoceran *Diaphanosoma aspinosum* in laboratory culture. 1988 35: 67-73.

Segawa, S. and W. T. Yang 1990. Growth, moult, reproduction, and filtering rate of an estuarine cladoceran, *Diaphanosoma celebensis* in laboratory culture. Bull. Plankton Soc. Japan 37: 145-155.

Segers, H. 1995. Nomenclatural consequences of some recent studies on *Brachionus plicatilis* (Rotifera, Brachionidae). Hydrobiologia 313/314: 121-122.

Segers, H. 2008. Global biodiversity of rotifers (Rotifera) in freshwater. Hydrobiologia 595:49-59.

Seikai, T., T. Takeuchi, and G. Park. 1997. Comparison of growth, feed efficiency, and chemical composition of juvenile flounder fed live mysids and formula feed under laboratory conditions. Fisheries Science 63:520-526.

Serra, M. and M. Miracle 1983. Biometric variation in three strains of *Brachionus plicatilis* as a direct response to abiotic variables. Hydrobiologia 147: 83-89.

Shirota, A. 1970. Studies on the mouth size of fish larvae. Bulletin Japanese Society Scientific Fisheries 36:353-368

Skjermo, J. and Vadstein, O. 1993. Characterization of the bacterial flora of mass cultivated *Brachionus plicatilis*. Hydrobiologia, 255/256: 185-191.

Smith, G. M. 1950. The Fresh-water Algae of the United States. 2nd ed. McGraw-Hill, New York. 719 pp.

Snell, T. W. 1991. Improving the design of mass culture systems for the rotifer, *Brachionus plicatilis*. In: Rotifer and Microalgae Culture Systems. Honolulu, HI, The Oceanic Institute. pgs. 61-71.

Snell, T. W. and B. C. Winkler 1984. Isozyme analysis of rotifer proteins. Biochem. Syst. Ecol. 12: 199-202.

Snell, T. W., B. D. Moffat, C. R. Janssen and G. Persoone 1991. Acute toxicity tests using rotifers. III. Effects of temperature, strain, and exposure time on the sensitivity of *Brachionus plicatilis*. Environ. Toxicol. Water Quality 6: 63-75.

Snell, T. W., B. D. Moffat, C. R. Janssen and G. Persoone 1991b. Acute toxity tests using rotifers. IV. Effects of cyst age, temperature, and salinity on the sensitivity of *Brachionus calyciflorus*. Ecotoxicol. Environ. Safety 21: 308-317.

Snell, T. W., B. H. Rosen, et al. 1990. Rotifer culture using a new dried microalgae product. Presented at World Aquaculture Society Meeting Halifax, Nova Scotia.

Snell, T.W. and F.H. Hoff 1989. Managing rotifer populations for maximum production. Presented at the World Aquaculture Society Meeting, Los Angeles.

Snell, T.W. and F.H. Hoff. 1985. The effect of environmental factors on resting egg production in the rotifer *Brachionus plicatilis*. Journal World Mariculture Society 16:484-497.

Snell, T.W. and F.H. Hoff 1988. Recent advances in rotifer culture. Aquaculture Magazine 14:41-45.

Snell, T.W. and G. Persoone. 1989a. Acute toxicity bioassays using rotifers. I. A test for brackish and marine environments with *Brachionus plicatilis*. Aquatic Toxicology 14:65-80.

Snell, T.W. and G. Persoone. 1989b. Acute toxicity bioassays using rotifers. II. A freshwater test with *Brachionus rubens*. Aquatic Toxicology 14:81-92.

Snell, T. W., J. Kubanek, W. Carter, A. B. Payne, J. Kim, M. K. Hicks, and C. P. Stelzer. 2006. A protein signal triggers sexual reproduction in *Brachionus plicatilis* (Rotifera). Mar. Biol. 149: 763-773.

Snell, T.W. and K. Carrillo. 1984. Body size variation among strains of the rotifer *Brachionus plicatilis*. Aquaculture 37:359-367.

Snell, T.W. 1987. Sex, population dynamics and resting egg production in rotifers. Hydrobiologia 144:105-111.

Snell, T.W., M. Childress, E. Boyer and F.H. Hoff. 1987. Assessing the status of rotifer mass cultures. Journal World Aquaculture Society 18:270-277.

Snell, T. W. and B. D. Moffat 1992. A two-day life cycle test with *Brachionus calyciflorus*. Environ. Toxicol. Chem. 11: 1249-1257.

Snell, T.W. and C.R. Janssen. 1995. Rotifers in ecotoxicology: a review. Hydrobiologia 313/314:231-247.

Snell, T. W., J. Kubanek, W. Carter, A. B. Payne, J. Kim, M. K. Hicks, and C. P. Stelzer. 2006. A protein signal triggers sexual reproduction in Brachionus plicatilis (Rotifera). Mar. Biol. 149: 763-773.

Sorgeloos, P. 1980. The use of brine shrimp *Artemia* in Aquaculture. In the Brine Shrimp *Artemia*, Vol. 3: 25-46. Pub. Universa Press, Belgium.

Sorgeloos, P., E. Bossuyt, E. Lavina, M. Baeza-Mesa, and G. Persoone. 1977. Decapsulation of *Artemia* cysts: a simple technique for the improvement of the use of brine shrimp in aquaculture. Aquaculture 12:311-315.

Sorgeloos, P., P. Lavens, P. Leger and W. Tackaert 1993. The use of *Artemia* in marine fish larviculture. TML Conference Proceedings 3: 73-86.

Sorgeloos, P., P. Lavens, P. Leger, W. Tackaert and D. Versichele 1986. Manual for the Culture and Use of Brine Shrimp *Artemia* in Aquaculture. Prepared for the Belgian Administration for Development Cooperation and the FAO of the United Nations, Artemia Reference Center, Belgium.

Spotte, S.H. 1971. Four general rules for use of live foods. Sea Scope, Vol. 2 #1. Aquarium Systems Inc. Mentor Ohio.

Spotte, S.H. 1979. Seawater Aquariums: The captive environment, Wiley-Interscience, New York, N.Y., 413p.

Stein, J.P. (editor). 1973. Handbook of Phycological Methods. Culture Methods and Growth Assessments. Cambridge University Press, New York.

Stein, J.P. 1981. Spatial and temporal distribution of zooplankton in a low-salinity Mississippi bayou system. Ph.D. diss. University of Mississippi, USA.

Stoecker, D. K. and J. M. Capuzzo 1990. Predation on protozoa: its importance to zooplankton. J. Plank. Res. 12: 891-908.

Stogstad, A., L. Granskog, D. Klaveness. 1987. Growth of freshwater ciliates offered planktonic algae as food. Journal of Plankton Research 9: 503-512.

Stottrup, J. G., K. Richardson, E. Kirkegaard and N. J. Pihl 1986. The cultivation of *Acartia tonsa* for use as a live food source for marine fish larvae. Aquaculture 52: 87-96.

Stottrup, J.G. and N.H. Norsker. 1997. Production and use of copepods in marine fish larviculture. Aquaculture 155:231-247.

Stottrup, J.G. 2000. The elusive copepods: their production and suitability in marine aquaculture. Aquaculture Research 31:703-711.

Suatoni, E., Vicario, S., Rice, S., Snell, T.W., Caccone, A., 2006. Phylogenetic and biogeographic patterns in the salt water rotifer *Brachionus plicatilis*. Molecular Phylogenetics & Evolution, doi:10.1016/j.ympev.2006.04.025.

Suchar, V.A. and P. Chigbu. 2006. The effects of algae species and densities on the population growth of the marine rotifer, Journal of Experimental Marine Biology and Ecology 337:96-102.

Sukenik, A. and R. Wahnon 1991. Biochemical quality of marine unicellular algae with special emphasis on lipid composition. I. *Isochrysis galbana*. Aquaculture 97: 61-72.

Sukenik, A., O. Zmora and Y. Carmeli 1993. Biochemical quality of marine unicellular algae with special emphasis on lipid composition. II. *Nannochloropsis* sp. Aquaculture 117: 313-326.

Szyper, J. P. 1989. Nutritional depletion of the aquaculture feed organisms Euterpina acutifrons, Artemia sp. and Brachionus plicatilis during starvation. J. World Aquacult. Soc. 20: 162-169.

Takano, H. 1971. Notes on the raising of an estuarine copepod *Gladioferens imparipes* Thomson. Bull. Tokai Reg. Fish. Res. Lab. 64:81-87.

Tamaru, C. S., R. Murashige, C.-S. Lee, H. Ako and V. Sato 1993. Rotifers fed various diets of baker's yeast and/or *Nannochloropsis oculata* and their effect on the growth and survival of striped mullet (*Mugil cephalus*) and milkfish (*Chanos chanos*) larvae. Aquaculture 110: 361-372.

Tandler, A. 1985. The rotifer *Brachionus plicatilis* as food for the larvae of the gilthead bream, *Sparus aurata*. Israel Journal Zoology 33:205.

Taniguchi, A. 1978. Reproduction and life histories of the tintinnid ciliates (Review). Bull. Plankton Soc. Japan. 25(2): 123-134..

Develop. Prog. (1987-88). Fl. Dept. Ag. and Consumer Ser., Div. of Marketing. Tallahassee, FL.

Theilacker, G.H. and A.S. Kimball. 1984. Comparative quality of rotifers and copepods as food for larval fishes. CalCOFI Report 25:80-86.

Theilacker, G.H. 1987. Feeding ecology and growth energetics of larval northern anchovy, *Engraulis mordax* . Fishery Bulletin U.S. 85:213-228.

Thompson, I., D. S. Jones and P. Dreibelbis. 1980. Annual internal growth banding and life history of the ocean quahog *Arctica islandica* (Mollusca: Bivalvia). Marine Biological, Vol. 57:25-34.

Tolbert, N.E. and L.P. Zill. 1956. Excretion of glycolic acid by algae during photosynthesis. Journal Biological Chemistry 222:895-906.

Toledo, J. D. and H. Kurokura 1990. Cryopreservation of euryhaline rotifer *Brachnious plicatilis* embryos. Aquaculture 91: 385-394.

Treece, G.D. and N. Wohlschlag, 1987. Raising food organisms for intensive larval culture: I. Algae culture. In: Manual of Red Drum Aquaculture. Texas Extension Ser., Corpus Christi, pp. III.6-23.

Trotta, P. 1983. An indoor solution for mass production of the marine rotifer *Brachionus plicatilis* Muller fed on the marine microalgae *Tetraselmis suecica* Butcher. Aquacultural Engineering 2:93-100.

Turk, P. E., M. E. Krejci and W. T. Yang 1982. A laboratory method for the culture of *Arcartia tonsa* (crustacea: copepods) using rice bran. Jour. of Aquaculture & Aquatic Sci. Vol III, No. 2:25-27.

USEPA. 1991. Methods for Measuring the Acute Toxicity of Effluents and Receiving Waters to Freshwater and Marine Organisms. Report EPA-600/4-90/027.

Ushiro, M., J.-P. Yu and A. Hino 1990. Energy feedback in the culture of the rotifer *Brachionus plicatilis*. Second Asian Fisheries Forum : 157-160.

Ushiro, M., S. Yamasaki and H. Hirata 1980. Examinations of bacteria as food for *Brachionus plicatilis* in culture. Min. Rev. Data File Fisheries Res. 1: 96-106.

Ushiro, M., S. Yamasaki and H. Hirata. 1980. Examination of bacteria as food for *Brachionus plicatilis* in culture. Min. Reviews Data File Fisheries Research 1:96-106.

Vaughan, David, L. Creswell and M. Pardee. 1988. A manual for farming the hard shell clam in Florida. Aqua. Market Harbor Branch Oceanographic Institute, Ft Pierce, FL.

van Rijn, J. 1996. The potential for integrated biological treatment systems in recirculating fish culture - a review. Aquaculture 139: 181-201.

Ventura, R.F. and E.M. Enderez. 1980. Preliminary studies on *Moina* sp. production in freshwater tanks. Aquaculture 21: 93-96.

Verpraet, R., M. Chair, P. Leger, H. Nelis, P. Sorgeloos and A. de Leenheer 1992. Live-food mediated drug delivery as a tool for disease treatment in larviculture. The enrichment of therapeutics in rotifers and *Artemia* nauplii. Aquacult. Engineering 11: 133-139.

Verpraet, R., M.Chair, P. Leger, H. Nelis, P. Sorgeloos and A. Deleenheer 1992. Live food medicated drug delivery as a tool for disease treatment in laviculture - The enrichment of therapeutics in rotifers and artemia nauplii. Aquaculture Eng. 11(2): 133-139.

Verschuere, L., G. Rombaut, P. Sorgeloos, and W. Verstraete. 2000. Probiotic bacteria as biological control agents in aquaculture. Microbiol. Mol. Biol. Rev. 64:655-675.

Walker, K.F. 1981. A synopsis of ecological information on the saline lake rotifer *Brachionus plicatilis* Muller 1786. Hydrobiologia 81:159-167.

Wallace, R.L. and T.W. Snell. 2001. Rotifera. In: Ecology and Systematics of North American Freshwater Invertebrates. Thorp, J.H. and A.P. Covich (eds.), Academic Press, NY., pp. 195-254.

Ward, S.H. 1984. A system for laboratory rearing of the mysid, *Mysidopsis bahia* Molenock. Progressive Fish Cult. 46:170-175.

Ward, S.H. 1987. Feeding response of the mysid *Mysidopsis bahia* reared on *Artemia*. Progressive Fish Cult. 49:29-33.

Ward, S.H. 1993. A comparison of natural and artificial seawater for culturing and toxicity testing with *Mysidopsis bahia*. In: W. Landis, J.S. Hughes, and M.A. Lewis, (eds.), Proceedings 1st ASTM Symp. Environ. Toxicol. & Risk Assessment, Philadelphia, PA.

Watanabe, K., K. Sezaki, K. Yazawa and A. Hino 1992. Nutritive fortification of the rotifer *Brachionus plicatilis* with eicosapentaenoic acid-producing bacteria. Nippon Suisan Gakkaishi 58: 271-276.

Watanabe, T. 1993. Importance of docosahexaenoic acid in marine larval fish. J. World Aquacult. Soc. 24: 152-161.

Watanabe, T., C. Kitajima and S. Fujita 1983a. Nutritional value of live food organisms used in Japan for mass culture of fish: A review. Aquaculture 34:115-143.

Watanabe, T., T. Tamiya, A. Oka, M. Hirata, C. Kitajima and S. Fujita 1983b. Improvement of dietary value of live foods for fish larvae by feeding them on w3 highly unsaturated fatty acids and fat-soluble vitamins. Bulletin Japanese Society Scientific Fisheries 49:471-479.

Yasuda,K. and N. Taga 1980. Culture of *Brachionus plicatilis* Muller using bacteria as food. Bulletin Japanese Society Scientific Fisheries 46:933-939.

Watanabe, T., T. Tamiya, A. Oka, M. Hirata, C. Kitajima, and S. Fujita. 1983. Improvement of dietary value of live foods for fish larvae by feeding them on omega-3 highly unsaturated fatty acids and fat-soluble vitamins. Bulletin Japanese Society Scientific Fisheries 49(3): 471-480.

Webster, K. and R. Peters. 1978. Some size-dependent inhibitions of larger cladoceran filterers in filamentous suspensions. Limnology and Oceanography 23(6): 1238-1245.

Whittingham, C.P. and G.G. Pritchard. 1963. The production of glycolate during photosynthesis in *Chlorella*. Proceedings Royal Society, B 157:366-380.

Whitford, L. A. and Schumacher, G. J. 1969. A Manual of the Fresh-water Algae of North Carolina. North Carolina Agric. Exp. Stat., Tech. Bull. No. 188. 313 pp.

Whyte, J. N. C., W. C. Clarke, N. G. Ginther, J. O. T. Jensen and L. D. Townsend 1994. Influence of composition of *Brachionus plicatilis* and *Artemia* on growth of larval sablefish (*Anoplopoma fimbria* Pallas). Aquaculture 119: 47-61.

Wickins, J.F., T.W. Beard, A.R. Child. 1995. Maximizing lobster, *Homarus gammarus*, (L.), egg and larval viability. Aquaculture Research 26:379-392.

Witt, U., G. Quantz, D. Kuhlmann, and G. Kattner., 1984. Survival and growth of turbot larvae *Scophthalmus maximus* L. reared on different food organisms with special regard to long-chain polyunsaturated fatty acids. Aquaculture Engineering, 3: 177-190.

Yamauchi, S. 1993. The effect of anitbacterial substances on the growth of the rotifer *Brachionus plicatilis*. Nippon Suisan Gakkaishi 59: 1001-1006.

Yasuda, K. and N. Taga 1980. Culture of *Brachionus plicatilis* using bacteria as food. Nippon Suisan Gakkaishi 46: 933-939.

Yoshimura, K., C. Kitajima, Y. Miyamoto and G. Kishimoto 1994. Factors inhibiting growth of the rotifer *Brachionus plicatilis* in high density cultivation by feeding condensed *Chlorella*. Nippon Suisan Gakkaishi 60: 207-213.

Yoshimura, K., T. Iwata, K. Tanaka, C. Kitajima and F. Isizaki 1995. A high density cultivation of rotifer in an acidified medium for reducing undissociated ammonia. Nippon Suisan Gakkaishi 61: 602-607.

Yoshimura, K., K. Usuki, T. Yoshimatsu, C. Kitajima, and A. Hagiwara. 1997. Recent development of a high density mass culture system for the rotifer *Brachionus rotundiformis*. Hydrobiologia, in press.

Yu, J.-P. and K. Hirayama 1986. The effect of un-ionized ammonia on the population growth of the rotifer in mass culture. Nippon Suisan Gakkaishi 52: 1509-1513.

Yu, J.-P., A. Hino, M. Ushiro and M. Maeda 1989. Function of bacteria as vitamin B_{12} producers during mass culture of the rotifer *Brachionus plicatilis*. Nippon Suisan Gakkaishi 55: 1799-1806.

Yu, J.-P., A. Hino, R. Hirano and K. Hirayma 1988. Vitamin B_{12}-producing bacteria as a nutritive complement for a culture of the rotifer *Brachionus plicatilis*. Nippon Suisan Gakkaishi 54: 1873-1880.

Yu, J.-P., A. Hino, T. Noguchi and H. Wakabayashi 1990. Toxicity of *Vibrio alginolyticus* on the survival of the rotifer *Brachionus plicatilis*. Nippon Suisan Gakkaishi 56: 1455-1460.

Yufera, M. and N. Navarro 1995. Population growth dynamics of the rotifer *Brachionus plicatilis* cultured in non-limiting food condition. Hydrobiologia 313/314: 399-405.

Yufera, M. 1982. Morphometric characterization of a small-sized strain of Brachionus plicatilis in culture. Aquaculture 27: 55-61.

Zerberj, W.B. and C. B. Taylor. 1953. Seawater temperature and density reduction tables, Special Publication.

Zillioux, E.J. 1969. A continuous recirculating culture system for plankton copepods. Marine Biology 4:215-218.

Zillioux E.J. and N.F. Lackie. 1970. Advances in the continuous culture of planktonic copepods. Helgolander Wiss. Meeresunters. 20:325-332

Zurlini, G., I. Ferrari and A. Nassogne. 1978. Reproduction and growth of *Euterpina acutifroms* (Copepoda: Harpacticoida) under experimental conditions. Marine Biology 46:59-64.

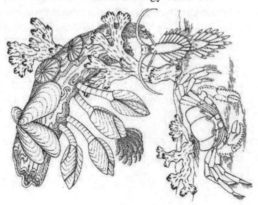

Dover Publications
Seashore Life 1973
Artist; Anthony D"Attilio

Metric Conversions

Cubic Foot (ft³)
28.32 L
28.317 ml or cc
7.48 gallons
62.43 lb of water
957.5 fl oz

Cup
8 fl oz
1/2 pt
237 ml

Gallon (gal)
3.75 L
3.785.4 ml or cc
128 fl oz
0.13 ft³
133.52 oz of water
8.35 lb of water

Gram (g)
0.035 oz
1 ml or cc of water
1,000 mg
1,000,000 µg

Liter (L)
33.82 fluid oz.
1.057 qt
0.26 gal
1 kg of water
2.20 lb of water
1000 ml or cc

Teaspoon (tsp)
4.93 ml or cc
1/3 tbsp
1/6 fl oz

Tablespoon (tbsp)
14.79 ml or cc
3 tsp
1/2 fl oz

Ounce (weight)
28.35 gm
0.063 lb
0.96 fl oz of water

Ounce (fluid)
29.57 gm of water
29.57 ml or cc
1.043 oz of water
1/8 cup
6 tsp
2 tbsp

Part per million
(ppm)
1 mg/L of water
3.78 mg/gal
3.78 g/1,000 gal
of water
0.13 oz/1,000 gal
of water
1 oz/1,000 ft³

Pint (pt)
473.17 ml or cc
1/2 qt
16 fl oz
1/8 gal
1.04 lb of water

Quart (qt)
946.34 ml or cc
32 fl oz
4 cups
2 pt
1/4 gal
2.09 lb

Meter (m)
39.37 in
100 cm
1000 mm
1,000,000 µm

Cubic Meter (m³)
35.31 ft³
1.308 yd³
264.2 gallons
1000 L
1234 acre-feet

Abbreviations in table above:

Cubic centimeter= cc
Cubic inch= in³
Cubic yard= yd³
Fluid Ounce= fl oz
Pound = lb
Ounce = oz
Meter= m
Milligram= mg
Milliliter= ml
Square meter= m²
Square yard= yd²
Square foot = ft²

Temperature Conversions

°C	°F
0	32.0
1	33.8
2	35.6
3	37.4
4	39.2
5	41.0
6	42.8
7	44.6
8	46.4
9	48.2
10	50.0
11	51.8
12	53.6
13	55.4
14	57.2
15	59.0
16	60.8
17	62.6
18	64.4
19	66.2
20	68.0
21	69.8
22	71.6
23	73.4
24	75.2
25	77.0
26	78.8
27	80.6
28	82.5
29	84.3
30	86.0
31	87.8
32	89.6
33	91.4
34	93.2
35	95.0
36	96.8
37	98.6
38	100.4
39	102.2
40	104
41	105.8
42	107.6

$$°C = \frac{°F - 32}{1.8}$$

$$°F = 1.8 \times °C + 32$$

Conversion of Salinities in ppt To Densities

Salinity	Density	Salinity	Density	Salinity	Density
1.1 ppt	1,0000	16.0 ppt	1.0114	31.0 ppt	1.0229
2.0	1.0007	17.0	1.0122	32.0	1.0237
3.0	1.0015	18.0	1.0130	33.1	1.0245
4.1	1.0023	19.0	1.0137	34.0	1.0252
5.0	1.0030	20.0	1.0145	35.0	1.0260
6.0	1.0038	21.0	1.0153	36.0	1.0268
7.1	1.0046	22.0	1.0160	37.1	1.0276
8.0	1.0053	23.0	1.0168	38.0	1.0283
9.0	1.0061	24.1	1.0176	39.0	1.0291
10.1	1.0069	25.0	1.0183	40.1	1.0299
11.0	1.0076	26.0	1.0191	41.0	1.0306
12.0	1.0084	27.1	1.0199	42.0	1.0316
13.1	1.0092	28.0	1.0206	43.0	1.0330
14.0	1.0099	29.0	1.0214		
15.0	1.0107	30.1	1.0222		

*Source Zerbe & Taylor (1953) Density taken at 15°F, Salinity parts per thousand parts of water

Major Ions in Seawater
(Salinity = 35 ppt)

ION	Chemical Name	g/kg^{-1}
Cl	Chloride	19.354
SO^2	Sulfate	2.712
Br	Bromine	0.0673
F	Iron	0.0013
B	Boran	0.0045
Na	Sodium	10.77
Mg^2	Mangnesium	1.290
Ca^2	Calcium	0.4121
K	Potassium	0.399
Sr^2	Strontium	0.0079

Ionic Composition of Seawater
(Salinty = 35 ppt)

Element	Chemical Species	Name	Molar	μg/L
H	H2O	Water	55	1.1x10^8
He	He (gas)	Helium	1.7x10^{-9}	6.8x10^{-3}
Li	Li$^+$	Lithium	2.6x10^{-5}	180
Be	BeOH$^+$	Berium	6.3x10^{-5}	5.6x10^{-3}
B	B(OH)$_3$, B(OH)$_4$	Boron	4.1x10^{-4}	4440
C	HCO$_3$, CO$_3$, CO$_2$	Carbon	2.3x10^{-3}	2.8x10^4
N	N^2, NO$_3$ NO$_2$, NH$_4$	Nitrogen	1.07x10^{-2}	1.5x
O	H2O, O2	Oxygen	55	8.8x108
F	F-, MgF+	Flourine	6.8x10-5	1.3x103
Ne	Ne (gas)	Neon	7x10-9	1.2x10-1
Na	Na+	Sodium	4.68x10-1	10.77x106
Mg	Mg2+	Magnesium	5.32x10-2	12.9x105
Al	Al(OH)-4	Aluminum	7.4x10-8	2
Si	Si(OH)4	Silica	7.1x10-5	2x106
P	HPO24-, PO34-, H2PO3	Potassium	2x10-6	60
S	SO24-, NaSO-4	Sulphur	2.82x10-2	9.05x105
Cl	Cl-	Chlorine	5.46x10-1	18.8x106
Ar	Ar (gas)	Argon	1.1x10-7	4.3
K	K+	Potassium	1.02x10-2	3.8x105
Ca	Ca2+	Calcium	1.02x10-2	4.12x105
Sc	Sc(OH)3		1.3x10-11	6x10-4
Ti	Ti(OH)4		2x10-8	1
V	H2VO4-, HVO24		5x10-8	2.5
Cr	Cr(OH)3, CrO24-		5.7x10-9	0.3

Oxygen Saturation ppm O$_2$

O$_2$ Saturation in Saltwater

°C	°F	20 ppt	30 ppt
10	50	9.9	9.0
15	59	8.9	8.1
20	68	8.1	7.4
22	72	7.8	7.1
24	75	7.5	6.9
26	79	7.2	6.6
28	83	7.0	6.4
30	86	6.8	6.2
35	95	6.6	5.8

O$_2$ Saturation in Freshwater

°C	°F	Sea Level	2,000 ft (600 M)
10	50	11.3	10.5
15	59	10.1	9.4
20	68	9.1	8.4
22	72	8.7	8.1
24	75	8.4	7.8
26	79	8.1	7.5
28	83	7.8	7.3
30	86	7.5	7.0

Amount of Chemical, at 1 ppm with 100% Active Ingredient

2.72 lbs/acre foot

1,233 gms/acre foot

0.0283 grams/cu foot

0.0000624 lbs/cu foot

0.0038 grams/gallon

0.00584 grains per gallon

1 milligram per liter

0.001 gram per liter

8.34 lbs/million gal of water

Conversion for ppm in Proporation & Percent

ppm	proportion	percent
0.1	1:10,000,000	0.000010
0.25	1:4,000,000	0.000025
0.5	1:2,000,000	0.00005
1.0	1:1,000,000	0.0001
2.0	1:500,000	0.0002
3.0	1:333,333	0.0003
4.0	1:250,000	0.0004
5.0	1:200,000	0.0005
8.4	1:119,047	0.0008
10.0	1:100,000	0.001
15.0	1:66,667	0.0015
20.0	1:50,000	0.002
25.0	1:40,000	0.0025
50.0	1:20,000	0.005
100.0	1:10,000	0.01
200.0	1:5,000	0.02
250.0	1:4,000	0.025
500.0	1:2,000	0.05

Estimation Of Fish Weight

(Number of fingerlings) X (Initial weight of fingerlings) = Initial total fingerling weight.

(Total pounds of feed fed) divided by 1.75 (approximate conversion ratio) = pounds of gain

(Initial total fingerling weight) + (pounds of gain) = Current weight of fish.

(Current weight of fish) divided by (Number of fish = Average current weight of fish.

Estimating Total Fish Biomass & Feed Levels

Fish Size	Avg. #/lb	Est Wt /K lbs	Protein	Feed Size	Est Lbs Feed/1K
1"	1,428.0	0.7 lbs	36%	1/16"	0.021/K
2"	322.6	3.1 lbs	36%	1/16"	0.093/K
3"	113.6	8.8 lbs	36%	1/8"	0.264/K
4"	52.3	19.1 lbs	36%	1/8"	0.573/K
5"	28.3	35.3 lbs	36%	3/16"	1.059/K
6"	17.0	58.8 lbs	36%	3/16"	1.764/K
7"	10.9	91.0 lbs	36%	3/16"	2.730/K
8"	7.5	133.3 lbs	32%	1/4"	3.999/K

Estimated quantity is based on water temperature of 75° F and a feeding rate of 3% of the biomass. Data based on catfish. (Gold Kist) Index: K = 1000

Calculating Amount of Chemical Needed

$$\text{Amt. Chem} = V \times C.F. \times ppm \times \frac{100}{\% A.I}$$

V = Determine the volume of the unit to be treated.

C.F. = Conversion factor is that weight of chemical used to give one ppm in one unit volume of water.(table left)

ppm = Concentration desired of chemical

$\frac{100}{\% A.I.}$ = 100 divided by the percent of active ingredient of the chemical.

Example: masoten 80% active needed 0.5 ppm active ingredient in a pond 100' x 100' x 4' deep.

V = L x W x Depth = cu ft of pond or tank

Divide by 43,560 cubic feet to convert to large numbers into acre-feet.

$$0.92 \times 2.72 \text{ lb} \times 0.5 \text{ ppm} \times \frac{100}{80\%} = 1.55 \text{ lbs}$$

If you use cu ft use .0000624 lb conversion number from table on left etc.

PRESSURE RELATIONSHIPS

Converting Water Pressure (PSI) to Feet Head

Pounds/IN²	Feet Head	Pounds/IN²	Feet Head
1	2.31	100	230.90
2	4.62	110	253.91
3	6.93	120	277.07
4	9.24	130	300.16
5	11.54	140	323.25
6	13.85	150	346.34
7	16.16	160	369.43
8	18.47	170	392.52
9	20.78	180	415.61
10	23.09	200	461.78
15	34.63	250	577.24
20	46.18	300	692.69
25	57.72	350	808.13
30	69.27	400	922.58
40	92.36	500	1154.48
50	115.45	600	1385.39
60	138.54	700	1616.30
70	161.63	800	1847.20

One Pound of pressure per square inch of water equals 2,309 feet of water at 62° Fahrenheit.

Converting Feet Head of Water to PSI

Feet Head	Pounds/IN²	Feet Head	Pounds/IN²
1	0.43	100	43.31
2	0.87	110	47.64
3	31.30	120	51.97
4	1.73	130	56.30
5	2.17	140	60.63
6	2.60	150	64.96
7	3.03	160	69.29
8	3.46	170	73.63
9	3.90	180	77.96
10	4.33	200	86.62
15	6.50	250	108.27
20	8.66	300	129.93
25	10.83	350	151.58
30	12.99	400	173.24
40	17.32	500	216.55
50	21.65	600	259.85
60	25.99	700	303.16
70	30.32	800	346.47
80	34.65	900	389.78

One pound of water at 62° Fahrenheit equals 0.433 pounds pressure per square inch.

Inches of Water (IN/H₂O) x 0.07353 = IN./Hg
Inches of Mercury (IN/Hg) x 13.6 = IN/H₂O
Inches of Water (IN/H₂O) x 0.036 = PSI
Feet of Water (FT/H2O) x 0.433 = PSI
Inches of Mercury (IN/Hg) x 0.490 = PSI

Centimeters of Mercury (Cm/Hg) x 0.193 = PSI
Millimeters of Mercury (Mm/Hg) x 0.193 = PSI
Microns of Mercury (μ/Hg) x 0.000019337 = IN/Hg
Microns of Mercury (μ/Hg) x 0.00003937 = IN/Hg

Conversion For Pressure Equivalents (Item X Number)

FROM to	POUNDS/IN²	FEET WATER	IN. MERCURY	ATMO-SPHERES
Pounds/IN²	1.0	2.309	2.036	0.068
Feet Water	0.433	1.0	0.882	0.0295
Inches Mercury	0.491	1.134	1.0	0.033
Atmospheres (atm)	14.7	33.9	29.92	1.0
Kilopascals (kPa)	0.145	0.335	0.295	0.010

Inches Mercury = a unit of pressure as measured by a manometer equal to the pressure balanced by the weight of a one inch column of mercury in the instrument.

Atmosphere = the weight of the atmosphere per square inch of surface; the pressure of 14.659 pounds per square inch exerted in all directions at sea level by the atmosphere.

Kilopascal = 1,000 pascals = a unit of force equal to one Newton per square meter (N/m²); typically used as pascal second (Pa - s), or 10 poise designating absolute viscosity.

Bar = a metric unit of measure often used, equal to 100 kilopascal (kPa) or 14.5 lb/in.²

STANDARD EVERYDAY CONVERSION INFORMATION

Temperature Conversions

F° = (9/5 x C°) + 32

C° = (F° x 5/9) - 32

Area Conversions & Equivalents

1 acre = 43,681 sq ft = 209 x 209

1 hectare = 2.47 acres

1 hectare = 10,000 sq. meters

1 sq meter = 10.76 sq. ft.

6.45 sq. cm = 1 sq. inch

1 sq mile = 640 acres

Volume Conversions & Equivalents

1 acre foot = 325,825 gallons

1 cubic foot = 7.48 gallons = 1728 cubic inches

1 cubic meter = 264.2 gallons

1 liter = 1.057 quarts = 33.82 fluid ounces

1 quart = 946 ml =4 cups

1 gallon = 3.78 liters = 128 fluid ounces = 231 cubic inches = 8.3453 pounds

1 liter = 1000 ml = 1000 cubic centimeters

1 fluid ounce = 29.57 ml = 29.57 cubic centimeters

3 teaspoons = 1 tablespoon - 14.79 ml

1 teaspoon = 4.93 ml

1/2 teaspoon = 2.47 ml

1/4 teaspoon = 1.23 ml

1 gallon = 231 cubic inches = 0.13 cubic feet

Linear Conversions & Equivalents

1 inch - 2.54 centimeters = 254 mm

1 meter = 39.4 inches = 3.28 feet

1 mile = 1.6 kilometers = 5280 feet

1 millmeter = 1000 microns = 0.0394 inches

1 rod = 16.5 feet

1 fathom = 6 feet = 1.83 meters

1 foot = 30.48 centemeters = 304.8 mm

Electical Conversion & Equivalents

Watt = 0.001 kilowatts

Watt - 0.001341 horsepower

Amps = watts/volts = 1 phase

Amps = watts/volts x 1.73 = 3 phase

Hp = 0.746 KW